International Cryogenics Monograph Series

The International Cryogenics Monograph Series was established in the early 1960s to present an opportunity for active researchers in various areas associated with cryogenic engineering to cover their area of expertise by thoroughly covering its past development and its present status. These high level reviews assist young researchers to initiate research programs of their own in these key areas of cryogenic engineering without an extensive search of literature.

More information about this series at http://www.springer.com/series/6086

Jacob W. Leachman · Richard T Jacobsen
Eric W. Lemmon · Steven G. Penoncello

Thermodynamic Properties of Cryogenic Fluids

Second Edition

Springer

Jacob W. Leachman
School of Mechanical and Materials Engineering
Washington State University
Pullman, WA
USA

Eric W. Lemmon
Applied Chemicals and Materials Division
National Institute of Standards and Technology
Boulder, CO
USA

Richard T Jacobsen
Center for Advanced Energy Studies
Idaho State University
Idaho Falls, ID
USA

Steven G. Penoncello
Department of Mechanical Engineering
University of Idaho
Moscow, ID
USA

ISSN 0538-7051 ISSN 2199-3084 (electronic)
International Cryogenics Monograph Series
ISBN 978-3-319-57833-0 ISBN 978-3-319-57835-4 (eBook)
DOI 10.1007/978-3-319-57835-4

Library of Congress Control Number: 2017938289

Printed on acid-free paper

This Springer imprint is published by Springer Nature
The registered company is Springer International Publishing AG
The registered company address is: Gewerbestrasse 11, 6330 Cham, Switzerland

Preface to the Second Edition

This edition continues as a source intended for the practicing engineer and interested scientist in accurate thermodynamic property information for cryogenic fluids. Cryogenic fluids display a wonderful spectrum of fluid behavior that spans the highly quantum regime near absolute zero through the classical regime of the very air we breathe. It is our hope that the reader finds enjoyment in visualizing how small changes in atomic and molecular structure result in significant changes to thermodynamic properties in the cryogenic regime.

In the 20 years since the first edition of Thermodynamic Properties of Cryogenic Fluids, the field of thermophysical properties is reaching maturity. Although sparse cryogenic experimental studies have been completed over this time, the computational fitting techniques have greatly improved, even though the functional form of pure fluid equations of state has essentially remained consistent. This maturation has allowed expansion of established techniques to historically challenging fluids and refinement of the software packages used to distribute this information.

The first edition of this text utilized a software package called ICMPROPS developed within the Center for Applied Thermodynamic Studies (CATS) at the University of Idaho. Shortly after the completion of the first edition, Eric Lemmon was employed by the National Institute of Standards and Technology (NIST) to incorporate ICMPROPS into the new software package REFPROP. REFPROP has superseded ICMPROPS and ALLPROPS and remains a leading software package for thermophysical property information. REFPROP can be downloaded directly from NIST, and a special version is available containing the fluids specific to this text. Additional information on REFPROP is included in this edition. As a result, this edition does not discuss the prior software packages.

This edition updates the majority of the property formulations as only neon, fluorine, oxygen, and methane remain unchanged. Xenon and ethane had the highest normal boiling points and were dropped from this edition corresponding to the established definition of the cryogenic fluid regime below 124 K. Work still remains, for example, the oxygen equation, now over 30 years old, is in the process of being refitted to change its use of the IPTS-68 to the ITS-90 scale—a change that alone will correct a 0.1% error in the vapor pressure.

The major change to this text is structural. Cryogenic applications typically utilize a single fluid. Therefore, this edition is formatted with each fluid comprising a chapter and arranged by normal boiling point, from the lowest (helium) through the highest (krypton). This approach allows an introduction to fluid-specific characteristics that may influence the implementation of the formulation. As with the first edition, we invite suggestions on the improvement and expansion of these tools by those who use them.

Pullman, WA, USA Jacob W. Leachman
Idaho Falls, ID, USA Richard T Jacobsen
Boulder, CO, USA Eric W. Lemmon
Moscow, ID, USA Steven G. Penoncello

Preface to the First Edition

This book is intended to provide the practicing engineer and interested scientist with the most accurate information available on the thermodynamic properties of cryogenic fluids. It may also be useful to universities and colleges as a supplementary reference text for elective courses in cryogenic engineering or engineering systems analysis that study systems with cryogens as the working fluids. Much of the material presented here is the result of a long-term continuing research effort in the Center for Applied Thermodynamic Studies (CATS) at the University of Idaho. While some of the thermodynamic property formulations presented here are the work of CATS staff, many have been developed and published by others. Numerical changes have been made to convert all of the correlations to a single form for ease of computation.

We have included the most accurate available formulation for each fluid, realizing that some of those included will be superseded in the future. We do not apologize for this circumstance, for it is the nature of this rapidly changing, dynamic field that both experimental methods and correlations improve with time. Fortunately, most new works extend the ranges of prior research or correct relatively minor errors in numerical models (e.g., near the critical point), and generally tend to verify the values given by models of the quality of those presented here.

In the presentation of correlations of thermodynamic data, it is customary to provide the reader or user with graphical and statistical information that verifies the accuracy of calculated properties. We have referenced the original works that contain these details for the interested user. We have provided sufficient detail on the model for each fluid that the user may program the formulations in any appropriate language or format consistent with a particular application.

In developing this book, we have given a brief review of the fundamentals of thermodynamic property formulations and a summary of current practices in data analysis and correlation. Although these discussions are intended to be very general, it is likely that the experiences of the authors have influenced the approach. The information included should be sufficient to allow the user to have confidence in the accuracy of calculated properties.

We have included Internet access to executable code for the computer programs used to calculate the fluid properties described in this book. It is our hope that the graphs, tables, and computer programs we have provided will be widely useful to those who need property information for cryogenic fluids. We invite suggestions on the improvement and expansion of these tools by those who use them.

Acknowledgements

This edition is dedicated to the lives and legacies of several cryogenic researchers that directly contributed to forming the first edition and have since passed. Dr. Richard B. Stewart made critical evaluations of the first edition, in addition to founding the Center for Applied Thermodynamic Studies at the University of Idaho, where much of this work began. Dr. Klaus Timmerhaus encouraged us to initiate this work in 1991. Dr. Anthony R.H. Goodwin assisted with the experimental uncertainties in the first edition. Their influence on this work cannot be overstated.

Timothy Rice and Agnes Baker at CryoGas International were very helpful by providing access to cryogenic fluid supply data utilized in the first chapter. Thomas Blackham contributed to initial formatting and structural changes of the second edition. The patience of the Springer staff in our drafting of this text was very appreciated.

Contents

1 Importance and Uses of Cryogenic Fluid Properties ... 1
1.1 Thermodynamic Property Information ... 2
1.2 Thermodynamic Property Formulations ... 3
1.3 Special Considerations for Modeling Cryogenic Fluids ... 6
1.4 Computer Programs for Thermodynamic Property Calculation ... 8
References ... 9

2 Equation of State Development and Use ... 11
2.1 Fundamental Equations of State ... 11
2.2 Development of Thermodynamic Property Formulations ... 13
2.2.1 Thermodynamic Property Data Correlation ... 13
2.2.2 Least-Squares Regression ... 14
2.2.3 Non-linear Least-Squares Regression ... 15
2.2.4 Criteria for Equations of State ... 16
2.2.5 Critical Region Behavior ... 17
2.3 Use of Fundamental Equations of State for Cryogenic Fluids ... 18
2.4 Ancillary Functions ... 18
References ... 20

3 Helium ... 23
3.1 Formulation for Helium ... 24
3.2 Property Plots for Helium ... 26
3.3 Thermodynamic Data Tables for Helium ... 29
References ... 39

4 Hydrogen ... 41
4.1 Formulation for Parahydrogen ... 43
4.2 Property Plots for Parahydrogen ... 45
4.3 Thermodynamic Data Tables for Parahydrogen ... 48

4.4 Formulation for Normal Hydrogen . . . 57
4.5 Property Plots for Normal Hydrogen . . . 58
4.6 Thermodynamic Property Tables for Normal Hydrogen . . . 61
4.7 Formulation for Orthohydrogen . . . 70
References . . . 71

5 Deuterium . . . 73
5.1 Formulation for Deuterium . . . 75
5.2 Property Plots for Deuterium . . . 76
5.3 Thermodynamic Data Tables for Deuterium . . . 79
References . . . 87

6 Neon . . . 89
6.1 Formulation for Neon . . . 89
6.2 Property Plots for Neon . . . 91
6.3 Thermodynamic Data Tables for Neon . . . 94
Reference . . . 102

7 Nitrogen . . . 103
7.1 Formulation for Nitrogen . . . 103
7.2 Property Plots for Nitrogen . . . 105
7.3 Thermodynamic Data Tables for Nitrogen . . . 108
Reference . . . 116

8 Air . . . 117
8.1 Formulation for Air . . . 117
8.2 Property Plots for Air . . . 119
8.3 Thermodynamic Data Tables for Air . . . 122
Reference . . . 130

9 Carbon Monoxide . . . 131
9.1 Formulation for Carbon Monoxide . . . 131
9.2 Property Plots for Carbon Monoxide . . . 133
9.3 Thermodynamic Data Tables for Carbon Monoxide . . . 136
Reference . . . 143

10 Fluorine . . . 145
10.1 Formulation for Fluorine . . . 145
10.2 Property Plots for Fluorine . . . 147
10.3 Thermodynamic Data Tables for Fluorine . . . 150
Reference . . . 156

11 Argon . . . 157
11.1 Formulation for Argon . . . 157
11.2 Property Plots for Argon . . . 159
11.3 Thermodynamic Data Tables for Argon . . . 162
Reference . . . 169

12 Oxygen 171
12.1 Formulation for Oxygen. 171
12.2 Property Plots for Oxygen 173
12.3 Thermodynamic Data Tables for Oxygen. 176
Reference 183

13 Methane. 185
13.1 Formulation for Methane 185
13.2 Property Plots for Methane 187
13.3 Thermodynamic Data Tables for Methane 190
Reference 197

14 Krypton. 199
14.1 Formulation for Krypton 199
14.2 Property Plots for Krypton. 201
14.3 Thermodynamic Data Tables for Krypton 204
Reference 210

Index 211

Symbols

a	Helmholtz energy, J/mol
B	Second virial coefficient, cm^3/mol
C	Third virial coefficient, cm^6/mol^2
C_p	Isobaric heat capacity, J/(mol K)
C_v	Isochoric heat capacity, J/(mol K)
g	Gibbs energy, J/mol
h	Enthalpy, J/mol
M	Molar mass, g/mol
N_A	Avogadro's number, mol^{-1}
p	Pressure, MPa
R	Molar gas constant, J/(mol K)
r	Radius of interaction, m
s	Entropy, J/(mol K)
T	Temperature, K
w	Speed of sound, m/s
Z	Compressibility factor
α	Reduced Helmholtz energy
ε	Fluid-specific parameter of the L-J potential, J/mol
δ	Reduced density
Λ	Quantum parameter
$\hbar$	Reduced Planck's constant, J s
ρ	Molar density, mol/dm^3
σ	Intermolecular radius at maximum well depth, m
τ	Inverse reduced temperature

Subscripts

c	Critical-point property
r	Reduced property

t Triple-point property
s Saturated-fluid property
o Dilute-gas property

Superscripts

* Quantum reduced property
' Saturated liquid property
" Saturated vapor property
r Real-fluid property
o Ideal-gas property

Chapter 1
Importance and Uses of Cryogenic Fluid Properties

Worldwide cryogenics markets continue to grow with over \$16.18 billion in equipment expenditures in 2016 and \$22.42 billion expected by 2021 [1]. In North America alone, the revenues generated from custody exchange of cryogenic fluids exceeded \$20 billion in 2014 [2]. As our worldwide economies continue to globalize, the importance of accurate thermophysical properties for engineering, safety, and custody exchange will continue to increase. Studies of the importance of accurate thermophysical properties are given by Zudkevitch [3], Zudkevitch and Gray [4], Sengers and Klein [5], Chappelear et al. [6] and Haynes et al. [7]. Table 1.1, modified from Haynes et al. [7], estimates the relative importance of accuracy of thermophysical properties in specific process areas. Haynes et al. [7] present a discussion including examples of specific processes and equipment that depend upon individual property information, including additional references to literature on this subject.

Although the use of automated software programs for providing this information has increased, the reliance on historical, quality, and often difficult experimental measurements should not be overlooked. The majority of these measurements remain historical, as reviewed elsewhere [8–10]. During the past several decades, international and national organizations including those listed below have played significant roles in the coordination of efforts to curate these measurements for the thermophysical property data needs of the engineering, scientific, and technological communities world-wide:

1. International Union of Pure and Applied Chemistry (IUPAC)
2. Committee on Data for Science and Technology (CODATA) of the International Council of Scientific Unions (ICSU)
3. National Institute of Standards and Technology (NIST) formerly National Bureau of Standards (NBS)
4. Gas Processors Association (GPA)
5. Gas Technology Institute (GTI), formerly Gas Research Institute (GRI), and Institute of Gas Technology (IGT)

J.W. Leachman et al., *Thermodynamic Properties of Cryogenic Fluids*,
International Cryogenics Monograph Series, DOI 10.1007/978-3-319-57835-4_1

Table 1.1 The relative importance of thermophysical property data in cryogenic processes[a]

Process	Type of property[b]				
	Pressure-density temperature	Phase equilibria	Calorimetric properties	Vapor pressure-critical properties	Transport properties
Distillation	2	1	2	1	2
Compression and expansion	1	1	1	1	2
Heat exchange	3	1	1	2	2
Metering and sales	1	2	2	2	3
Storage	2	1	1	2	3
Safety	2	1	1	2	3
Alternative fuels	1	2	1	2	3

[a]Modified from Haynes et al. [7]
[b]1 = most important, 2 = very important, 3 = less important

6. American Society of Heating, Refrigerating, and Air-Conditioning Engineers, Inc. (ASHRAE)
7. American Society of Mechanical Engineers (ASME)
8. American Institute of Chemical Engineers (AIChE)
9. Japanese Association of Refrigeration (JAR)
10. International Energy Agency (IEA)

1.1 Thermodynamic Property Information

Modern thermodynamic property formulations are developed by considering experimental measurements, theoretical property estimates, and molecular simulation results. The importance of accurate experimental data in the development of a thermodynamic property correlation is well established. Systematic errors in experimental data sets affect the quality of the correlation, which is based upon such data. To minimize the effects of such errors, care is taken to analyze each data set used in a correlation for consistency with independent measurements and with data forms in the same and adjacent regions of the thermodynamic surface. It is best to avoid the use of experimental data that contain large systematic errors. Large random errors may be tolerated when no alternative data sets are available, but the use of such data increases the uncertainty of the formulation. If data are not available for a particular region of the thermodynamic surface, it is often helpful to use predicted values and allow constraints in the fitting process to ensure correct behavior (see Sect. 2.2.3). Property formulations utilize a fitting process that allows preferential weights on the data to allow accurate predictions while maintaining behavior consistent with the laws of thermodynamics.

There are several fixed points and reference state properties that must be available for each fluid for which a thermodynamic property correlation is developed. These include:

1. Critical point temperature and density
2. Triple point temperature
3. Gas constant and molar mass

In addition, units and scales may be outdated, and procedures for converting temperatures measured on various scales to the currently accepted temperature scale to allow data from various sources to be converted to consistent units for analysis must be available. The following experimental data are essential in developing accurate property formulations:

1. Vapor pressure measurements
2. Ideal gas heat capacities over the range of temperature of the formulation
3. P-ρ-T data in all regions of the thermodynamic surface, including either saturated liquid density measurements or densities at 1 bar
4. W-P-T data (sound speed)

The accuracy of properties derived from a property formulation can be improved if the following data are also available:

5. C_p-P-T data (isobaric heat capacity)
6. C_v-ρ-T data (isochoric heat capacity)
7. Second and third virial coefficients
8. Joule-Thomson coefficient data

1.2 Thermodynamic Property Formulations

The term "thermodynamic property formulation" is used to define a group of equations required for the complete definition of the thermodynamic properties of a fluid. The equation of state is the major part of the formulation and generally requires the most data for its development.

Accuracy and thermodynamic consistency in a wide-range thermodynamic equation of state for a pure substance imply that all thermodynamic properties may be calculated within the limits of experimental uncertainty. There are regions where both experimental data and the correlation exhibit limitations. The critical region behavior should be consistent with experimental measurements and theoretical considerations (heat capacity diverges and speed of sound approaches zero at the critical point). However analytic equations cannot represent the singularities at the critical point. Specific criteria for equations of state are given in Chap. 2.

One desirable attribute of an equation of state for engineering applications is relative simplicity to provide for minimal computation time for the calculation of properties. However, a relatively large number of terms in the functional form for an equation of state may be required to accurately represent the thermodynamic properties of a fluid. If accuracy requirements can be relaxed, or for limited ranges of application, shorter equations of state may be developed to reduce computation times.

An equation of state represents the dependent variable as a function of n independent variables; that is, any dependent property y is a single-valued function of n independent properties x_1, x_2,..., x_n, and may be expressed as

$$y = f(x_1, x_2, \ldots, x_n) \tag{1.1}$$

Although equations of state are primarily developed with experimental measurements, guidance for behavior of fluids may be obtained from theoretical approaches that include statistical mechanics, and most recently, molecular simulation.

The simplest fluid systems in thermodynamics are pure homogeneous substances characterized by two independent properties. The most common measured properties used in the equation of state are pressure, temperature, and specific volume (or its reciprocal, density). A convenient equation of state may thus be expressed as

$$P = P(\rho, T) \tag{1.2}$$

However, the choice of the dependent variable is arbitrary. Needs for computational speed have necessitated modern equations of state to be explicit in a fundamental property, for instance internal energy, enthalpy, Gibbs energy, or Helmholtz energy. All other thermodynamic properties can be determined from derivatives of a fundamental property and hence avoid cumbersome numerical integration. Of the four fundamental properties, the Gibbs and Helmholtz energy have a combination of pressure and temperature or density and temperature, respectively, as inputs. Due to the needs to predict property changes through phase change, the Helmholtz energy a has emerged as the most commonly employed equation of state,

$$a = a(\rho, T). \tag{1.3}$$

This form is generally called a fundamental equation to indicate that it contains both calorimetric and compressibility information about the fluid in a single functional form.

Current practice in the development of computer programs, property tables, and charts for system analysis involves the correlation of selected experimental data for a particular fluid or mixture. The development of an accurate property formulation requires extensive analysis of the available data and correlation with an equation of state with a suitable functional form. The process of determining the optimum equation of state often involves art as well as science, and experience plays a part in the determination of the final result. The objective of the correlator is to ascertain the accuracy of the available experimental data for the particular fluid or system under investigation, and to develop a mathematical model capable of representing the data within the reported or estimated experimental uncertainty. When many (tens to hundreds) of overlapping experimental data sets are available for a fluid, this somewhat arbitrary process of assessing experimental consistency between datasets can be automated. However, as in the case with most fluid cryogens, limited experimental measurements are available and selection of primary versus secondary data sets is typically based on the expertise of the correlator. Compromises are often required and the limitations of the model selected should be understood by the user.

Some frustration may occur with the user of published property formulations when a seemingly accurate equation of state is replaced in the literature or property packages. In some cases, new formulations are done to improve the accuracy of older models that have been superseded by new correlations developed with improved techniques. The advantage to the user of such information is improved accuracy of engineering design or analysis from property formulations for cryogenic fluids.

The accuracy of properties calculated with computer programs based upon thermodynamic correlations is limited by the availability of experimental data to define the fluid behavior in various regions of the thermodynamic surface. For comparisons of calculated values to measured properties, the single-phase thermodynamic regions shown in Fig. 1.1 are arbitrarily defined as:

1. Liquid below the critical temperature
2. Vapor below the critical temperature

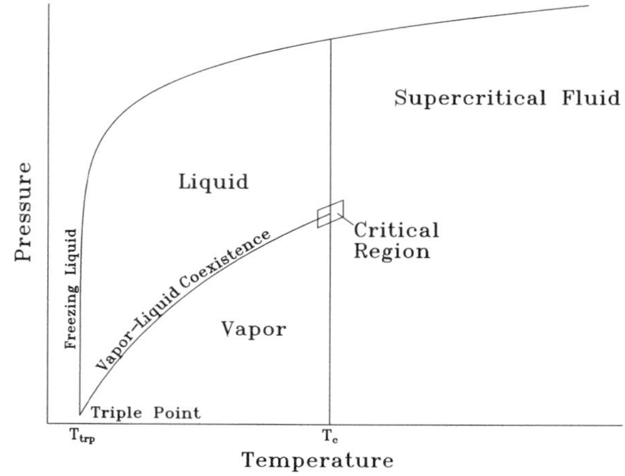

Fig. 1.1 Regions of the thermodynamic surface

Table 1.2 Estimated accuracies of various properties

Calculated property		State-of-the-art experimental accuracy	Accuracy to be expected from a typical equationof state (%)
Density		±0.02%	±0.1
Pressure	$P < 0.5$ MPa $P > 0.5$ MPa	±0.02% ±0.05%	
Temperature		±1 mK	
Isochoric heat capacity	$\rho > \rho_c$ $\rho < \rho_c$	±0.5% ±1%	±0.5 ±1
Isobaric heat capacity	$\rho > \rho_c$ $\rho < \rho_c$	±0.5% ±1%	±1 ±1
Speed of sound	$\rho > \rho_c$ $\rho < \rho_c$	±0.02% ±0.001%	±0.1 ±0.1
Vapor pressure	$P < 0.1$ MPa $P > 0.1$ MPa	±0.05% ±0.01%	±0.5 ±0.1

3. Fluid above the critical temperature
4. Critical region within 5% of the critical temperature and 25% of the critical density

The accuracies of calculated values of various properties are determined by comparing them to measured values. Graphs that illustrate such comparisons show differences in experimental data from alternate sources and facilitate the determination of random errors and relative consistency of data. Statistical analysis based on the percent deviation is also used to determine the overall estimated accuracy of various data sets that are selected for determining the coefficients of equations of state, and to define the accuracies for various properties calculated with the formulation.

The percent deviation for any property, X, is given by

$$\%\Delta X = 100\left(\frac{X_{\text{data}} - X_{\text{calc}}}{X_{\text{data}}}\right) \tag{1.4}$$

General estimates of the attainable accuracies of various properties are given in Table 1.2.

1.3 Special Considerations for Modeling Cryogenic Fluids

The fluids included in this book are those that have a normal boiling point temperature below 124 K as this value excludes most traditional refrigerants. Although air is clearly a mixture, it is included as a pseudo-pure fluid for convenience. The previous edition included property formulations for ethane and xenon, but which

Table 1.3 Cryogenic fluids included in this work and information on their equation of state (in order of normal boiling points)

Fluid	Reference	Temperature range (K)	Max. pressure (MPa)	Normal boiling point temperature (K)
Helium	Ortiz-Vega et al. [11]	2.1768–2000	1000	4.222
Hydrogen	Leachman et al. [12]	13.957–1000	2000	20.369
Deuterium	Richardson et al. [13]	18.71–600	2000	23.647
Neon	Katti et al. [14]	24.556–700	700	27.104
Nitrogen	Span et al. [15]	63.151–2000	2200	77.355
Air	Lemmon et al. [16]	59.75–2000	2000	78.903
Carbon Monoxide	Lemmon and Span [17]	68.16–500	100	81.64
Fluorine	de Reuck et al. [18]	53.481–300	20	85.037
Argon	Tegeler et al. [19]	83.806–2000	1000	87.302
Oxygen	Schmidt and Wagner [20]	54.361–2000	82	90.188
Methane	Setzmann and Wagner [21]	90.694–625	1000	111.67
Krypton	Lemmon and Span [17]	115.78–750	200	119.73

have been omitted from this version consistent with the new definition of the cryogenic temperature regime. Table 1.3 lists the fluids included, the source of the equation of state, and relevant state properties.

Fluid cryogens are generally considered the most ideal of all gases near ambient conditions. However, caution should be exercised when utilizing traditional corresponding states and other scaling techniques due to the large effect of quantum mechanics on thermophysical properties at cryogenic temperatures. In some cases, scaling techniques are required to estimate fluid properties where formulations are unavailable. Examples include rare species, such as helium-3 or tritium, and where melting or transport property data are unavailable.

The Quantum parameter is a useful comparison of traditional and quantum fluid behavior:

$$\Lambda^{*} = \frac{N_A 2\pi\hbar}{\sigma(M\varepsilon)^{1/2}}, \tag{1.5}$$

with Avogadro's number N_A, the reduced Planck's constant $\hbar$, the molar mass M, the maximum well depth from the Lennard-Jones potential ε, and the inter-particle radius of the maximum well depth σ. The Quantum parameter can be utilized to scale thermophysical properties between helium-4, hydrogen, deuterium, and neon. Once the quantum parameter is below 0.25, traditional corresponding states methods are applicable. In this form the Quantum parameter is a dimensionless comparison between the thermal de Broglie wavelength and average radius between particles. Values for the Quantum parameter for the fluids included in this work are shown in Table 1.4.

Table 1.4 Quantum parameters for cryogenic fluids

Fluid	Quantum parameter, Λ^*
Helium-4	2.67
Hydrogen	1.73
Deuterium	1.22
Neon	0.58
Nitrogen	0.226
Carbon monoxide	0.22
Fluorine	0.179
Argon	0.186
Oxygen	0.201
Methane	0.239
Krypton	0.102

The relatively strong quantum mechanical behavior of hydrogenic and helium particles raises additional considerations specific to each fluid. For example, the presence of a superfluid helium phase is unique among fluids. Other examples are hydrogen and deuterium that have independent statistical regimes due to parity with the nuclear spin quantum number that result in complicated calorimetric property behavior. As a result of these nuances, we have chosen to have a chapter specific to each fluid for explanation to aid with correct use of the formulations.

1.4 Computer Programs for Thermodynamic Property Calculation

The computer program for the calculation of thermodynamic properties of cryogens is available as a subset of the programs in REFPROP [22] provided by the National Institute of Standards and Technology (NIST). The REFPROP software contains equations of state for 121 fluids, including all of the most commonly used cryogens, hydrocarbons, and refrigerants in industry. This program calculates the properties of both single-phase and two-phase states, for all liquid, vapor, and supercritical states, and with upper temperature and pressure limits of at least 500 K and 50 MPa, except for fluids that dissociate at lower temperatures. For well measured substances, the limits are often 1000 K or more and 1 GPa or more. Both thermodynamic and transport properties are available for most fluids. The program comes with the FORTRAN source code used to calculate the properties, a graphical interface, and methods to link with most applications that can connect to a DLL, such as Excel or MATLAB. Mixture properties can be calculated in both the single-phase and two-phase regions, except for some mixtures with water, helium, and hydrogen that have not yet been fitted to experimental data to obtain the necessary interaction parameters required for such complex mixtures [due to either association (i.e., water) or the low boiling point of the fluid (i.e., helium or hydrogen) mixed with other high-boiling components].

A subset of this program is available free of charge from NIST, and is available at: http://trc.nist.gov/refprop/MINIREF/MINIREF.HTM.

The program contains only a limited number of fluids and does not allow the calculation of mixture properties. However, all other pure fluid calculations are available and work identically to those in the REFPROP program. The graphical interface allows the user to calculate properties, either one point at a time by entering two properties from any combination of temperature, pressure, density, internal energy, enthalpy, or entropy, or as a table of properties calculated along an isotherm, isobar, isochore (constant density), isenthalp (constant enthalpy), or isentrop (constant entropy). Tables can also be calculated along the saturation lines, or along lines of constant quality in the two-phase region. Plots can be generated through either the use of a predefined set of limits and constant properties lines, or through the setup of inputs requested by the user. Calculations from tables can be overlaid on existing plots, or used to generate new plots. These can all be saved in various formats to one's computer, or printed for use elsewhere.

References

1. Markets and Markets, Cryogenic equipment market by product (tank, valve, vaporizer, pump), cryogen (nitrogen, oxygen, argon, LNG), end-user (energy & power, chemicals, metallurgy, electronics, shipping) and region-global forecast to 2021, EP 3337 (Published September 2016), http://www.marketsandmarkets.com/Market-Reports/cryogenic-equipment-market-735.html
2. Gasworld, Gas reports: North America, utilizes 2014 as base year. https://www.gasworld.com/intelligence/gas-reports/north-america. Accessed December 2016
3. D. Zudkevitch, Imprecise data impacts plant design and operation. Hydrocarbon. Process. **54**, 97 (1975)
4. D. Zudkevitch, R.D. Gray Jr., Impact of fluid properties on the design of equipment for handling LNG, Adv. Cryo. Eng. **20** (Plenum Press, New York, 1975)
5. J.V. Sengers, M. Klein (ed.), *The Technological Importance of Accurate Thermophysical Property Information*, Special Publication 590, Nat. Bur. Stand. (U.S.) (1980)
6. P.S. Chappelear, R.J.J. Chen, D.G. Elliot, Pick K correlations carefully. Hydrocarbon. Process. **56**, 215 (1977)
7. W.M. Haynes, A.J. Kidnay, N.A. Olien, M.J. Hiza, Status of thermophysical properties data for pure fluids and mixtures of cryogenic interest. Adv. Cryo. Eng. **27**, 919 (1984)
8. F. Din, Thermodynamic functions of gases; Volume 1. Ammonia, Carbon Dioxide, and Carbon Monoxide; Volume 2. Air, Acetylene, Ethylene, Propane and Argon; Volume 3. Ethane, Methane, and Nitrogen (Butterworths, London, 1962)
9. R.B. Stewart, K.D. Timmerhaus, The correlation of thermodynamic properties of cryogenic fluids. Adv. Cryo. Eng. **9**, 20 (1964)
10. D.E. Diller, Thermophysical properties data research on compressed and liquefied gases at the NBS cryogenics division. Adv. Cryo. Eng. **21**, 522–531 (1976)
11. D.O. Ortiz-Vega, K.R. Hall, J.C. Holste, V.D. Arp, E.W. Lemmon, To be submitted to: J. Phys. Chem. Ref. Data (2013)

12. J.W. Leachman, R.T. Jacobsen, S.G. Penoncello, E.W. Lemmon, Fundamental equations of state for parahydrogen, normal hydrogen, and orthohydrogen. J. Phys. Chem. Ref. Data **38**(3), 721–748 (2009)
13. I.A. Richardson, J.W. Leachman, E.W. Lemmon, Fundamental equation of state for deuterium. J. Phys. Chem. Ref. Data **43**(1), 013103–1 (2014)
14. R.S. Katti, R.T. Jacobsen, R.B. Stewart, M. Jahangiri, Thermodynamic properties for neon for temperatures from the triple point to 700 K at pressures to 700 MPa. Adv. Cryog. Eng. **31**, 1189–1197 (1986)
15. R. Span, E.W. Lemmon, R.T. Jacobsen, W. Wagner, A. Yokozeki, A reference equation of state for the thermodynamic properties of nitrogen for temperatures from 63.151 to 1000 K and pressures to 2200 MPa. J. Phys. Chem. Ref. Data **29**(6), 1361–1433 (2000)
16. E.W. Lemmon, R.T. Jacobsen, S.G. Penoncello, D.G. Friend, Thermodynamic properties of air and mixtures of nitrogen, argon, and oxygen from 60 to 2000 K at pressures to 2000 MPa. J. Phys. Chem. Ref. Data **29**(3), 331–385 (2000)
17. E.W. Lemmon, R. Span, Short fundamental equations of state for 20 industrial fluids. J. Chem. Eng. Data **51**, 785–850 (2006)
18. K.M. de Reuck, *International Thermodynamic Tables of the Fluid State-11 Fluorine* (International Union of Pure and Applied Chemistry, Pergamon Press, Oxford, 1990)
19. Ch. Tegeler, R. Span, W. Wagner, A new equation of state for argon covering the fluid region for temperatures from the melting line to 700 K at pressures up to 1000 MPa. J. Phys. Chem. Ref. Data **28**(3), 779–850 (1999)
20. R. Schmidt, W. Wagner, A new form of the equation of state for pure substances and its applications to oxygen. Fluid Phase Equilib. **19**, 175 (1985)
21. U. Setzmann, W. Wagner, A new equation of state and tables of thermodynamic properties for methane covering the range from the melting line to 625 K at pressures up to 1000 MPa. J. Phys. Chem. Ref. Data **20**(6), 1061 (1991)
22. E.W. Lemmon, M. L. Huber, M.O. McLinden, *Standard Reference Data Program #23, REFPROP Version 9.1.* (National Institute of Standards and Technology, Gaithersburg, Maryland 2013)

Chapter 2
Equation of State Development and Use

Both historical and current functional forms for equations of state are described in the first edition of this book and other publications [1], and are not included here. Since the writing of the first edition, fundamental equations have become the standard source for accurate thermodynamic property information, and limited details are given in this section. The equations of state presented here all now use the same functional form, having replaced all historical forms.

2.1 Fundamental Equations of State

The term "fundamental equation" is generally reserved for those equations that contain calorimetric and datum state information so that absolute values of specified properties may be calculated directly by mathematical differentiation of a thermodynamic potential. The fundamental equations reported in this work are explicit in the Helmholtz energy a, which is the common form used for thermodynamic property formulations.

The use of fundamental equations of state in thermodynamic property correlations has several advantages over other forms:

1. The intrinsic advantage of a fundamental equation is that all properties may be calculated by differentiation. The typical pressure explicit equation of state must be integrated, for example, for the calculation of enthalpy and entropy.
2. The terms in the equation used for linear-fitting of property data may be changed by altering elements of an array defining the various coefficients and exponents. Section 2.2.2 gives details of the least-squares fitting process.
3. The temperature exponents can be fitted with the use of non-linear methods; further information on non-linear fitting is given in Sect. 2.2.3.
4. The form allows the correlator considerable flexibility in the design of an equation for a specific fluid.

J.W. Leachman et al., *Thermodynamic Properties of Cryogenic Fluids*,
International Cryogenics Monograph Series, DOI 10.1007/978-3-319-57835-4_2

5. Computer programs for fitting and for the calculation of thermodynamic property tables may be standardized so that few programming changes are required to calculate properties of several different fluids.

Schmidt and Wagner [2] developed a 32-term equation of state in 1985. Although this form was developed for oxygen, it has been used by other investigators for correlating properties of other fluids. The equation is explicit in dimensionless Helmholtz energy,

$$a(\rho, T) = a^0(\rho, T) + a^r(\rho, T) \tag{2.1}$$

where $a^0(\rho, T)$ is the ideal gas contribution to the Helmholtz energy and the term $a^r(\rho, T)$ represents the contribution due to the compressibility of the fluid. The pressure derived from this expression is

$$P = \rho^2 \left(\frac{\partial a}{\partial \rho}\right)_T \tag{2.2}$$

The Helmholtz energy for the ideal gas is given by

$$a^0 = h^0 - RT - Ts^0 \tag{2.3}$$

Combining expressions for h^0 and s^0, the Helmholtz energy of the ideal gas is given by

$$a^0 = h_0^0 + \int_{T_0}^{T} c_p^0 \mathrm{d}T - RT - T\left[s_0^0 + \int_{T_0}^{T} \frac{c_p^0}{T} \mathrm{d}T - R \ln\left(\frac{\rho T}{\rho_0 T_0}\right)\right] \tag{2.4}$$

An equation for the ideal-gas heat capacity as a function of temperature must be used with this equation to develop the complete expression for a specific fluid:

$$\frac{c_p^0}{R} = a_0 + \sum_{i=3}^{m} a_i T^{k_i} + \sum_{i=100}^{n} a_i \frac{(k_i/T)^2 \exp(k_i/T)}{[\exp(k_i/T) - 1]^2} \tag{2.5}$$

Practical equations of state generally use reduced variables to simplify computer calculations. Generally, properties are reduced by accepted critical state values or other combinations of properties and constants to make the equations nondimensional. The nondimensional variables used in this work are

$$\alpha = \frac{a}{RT}, \quad \delta = \frac{\rho}{\rho_c}, \text{ and } \tau = \frac{T_c}{T} \tag{2.6}$$

With these variables, Eq. (2.1) is written as

$$\alpha(\delta,\tau) = \alpha^0(\delta,\tau) + \alpha^r(\delta,\tau) \tag{2.7}$$

The ideal-gas contribution to the dimensionless Helmholtz energy can be derived from Eq. (2.4) as

$$\alpha^0 = \frac{h_0^0\tau}{RT_c} - \frac{s_0^0}{R} - 1 + \ln\frac{\delta\tau_0}{\delta_0\tau} - \frac{\tau}{R}\int_{\tau_0}^{\tau}\frac{c_p^0}{\tau^2}\,\mathrm{d}\tau + \frac{1}{R}\int_{\tau_0}^{\tau}\frac{c_p^0}{\tau}\,\mathrm{d}\tau \tag{2.8}$$

A computationally convenient parameterized form of Eq. (2.8) is

$$\begin{aligned}\alpha^0 &= \ln\delta + (a_0 - 1)\ln\tau + a_1 + a_2\tau - \sum_{i=3}^{m} a_i[T_c/\tau]^{k_i}/[k_i(k_i+1)] \\ &\quad + \sum_{i=m+1}^{n} a_i \ln[1 - \exp(-k_i\tau/T_c)].\end{aligned} \tag{2.9}$$

The fluorine formulation is the one exception to this equation and replaces T_c/τ with $1/\tau$. The form of $\alpha^r(\delta,\tau)$ is based on theoretical and practical considerations required to represent experimental data for a fluid. The general form used in current practice is

$$\begin{aligned}\alpha^r(\tau,\delta) &= \sum_{i=1}^{l} N_i\delta^{d_i}\tau^{t_i} + \sum_{i=l+1}^{m} N_i\delta^{d_i}\tau^{t_i}\exp(-\zeta_i\delta^{p_i}) \\ &\quad + \sum_{i=m+1}^{n} N_i\delta^{d_i}\tau^{t_i}\exp\left[-\phi_i(\delta - D_i)^2 - \beta_i(\tau-\gamma_i)^2\right]\end{aligned} \tag{2.10}$$

The parameters in these terms are determined through fitting experimental data. The terms in the third summation are usually described as Gaussian bell-shaped terms. The term ζ_i is only used in the EOS of fluorine and has a value of one for all other fluids.

2.2 Development of Thermodynamic Property Formulations

2.2.1 Thermodynamic Property Data Correlation

As established in Chap. 1, equations of state rely extensively on experimental measurements, theoretical estimates, and molecular simulation. The data selection process is a matter of experience and requires several tools for comparison of data sets, including:

1. Graphical displays of data (data maps) to assist in identifying regions of overlapping data from alternative sources.
2. Graphical plots that show deviations between experimental data and previously accepted correlations or to preliminary correlations.
3. Property algorithms for producing tables and charts that can be used to examine the behavior of properties that were not included in the correlation.

Contributions of individual data points are weighted before they are used in least-squares fitting. Weighting is based on estimated accuracy. Some form of the theorem of propagation of variance should be used to correct for different levels of random errors in data. Often additional weighting factors are applied to achieve a desired emphasis of one form or body of data over another.

2.2.2 Least-Squares Regression

Methods for performing simultaneous regression on multiple-property data are given by Hust and McCarty [3]. In the regression process, thermodynamic property relations are written to form fit variables to represent experimental data. The dimensionless fit variables listed in Table 2.1 are examples of such variables used in developing equations of state. In forming fit variables, it is necessary to "linearize" some data forms with the use of a preliminary equation of state. For example, velocity of sound data are often prepared for least-squares regression through the use of a density and a ratio of heat capacities calculated from a preliminary equation of state.

The objective of linear least-squares regression is to select the coefficients N_k that minimize the sum of squares of differences between each experimental fit variable

Table 2.1 Functions for fitting equations of state to various data forms

Data form	Experimental fit variable	Calculated fit variable
P-ρ-T	$F_P = \frac{P}{P_c} - \frac{\delta}{\tau Z_c}$	$F'_P = \frac{\delta^2}{\tau Z_c} \sum_{q=1}^{M} \left(\frac{\partial \alpha_i^r}{\partial \delta}\right)_\tau$
C_v-ρ-T	$F_{C_v} = \frac{C_v}{R} + \tau^2 \frac{\partial^2 \alpha^0}{\partial \tau^2}$	$F'_{C_v} = -\tau^2 \sum_{q=1}^{M} \left(\frac{\partial^2 \alpha^r}{\partial \tau^2}\right)_\delta$
W-ρ-T	$F_W = \left(\frac{w^2 \tau}{\gamma R T_c}\right) - 1$	$F'_W = \sum_{q=1}^{M} \left[2\delta \left(\frac{\partial \alpha^r}{\partial \delta}\right)_\tau + \delta^2 \left(\frac{\partial^2 \alpha^r}{\partial \delta^2}\right)_\tau\right]$
Liquid-vapor phase boundary	$F_\sigma = (Z_{S_L} - Z_{S_V}) + \ln(\delta_{S_L}/\delta_{S_V})$	$F'_\sigma = \alpha^r_{S_v} - \alpha^r_{S_L}$
B-T	$F_B = B\ \rho_c$	$F'_B = \sum_{q=1}^{M} \left[\sum_{k=1}^{m} \tau^{j_k}\right]_{i_k=1}$

where: σ = Saturation state; S_V = Saturated vapor; S_L = Saturated liquid; γ = Ratio of heat capacities (C_p/C_v); M = Number of data points; and i_k, j_k, and m come from Eq. (2.10)

F and the equivalent value calculated from the equation of state F'. The coefficients N_k are computed to minimize the sum of squares of the deviations S,

$$S = \sum Y_p (F_P - F'_P)^2 + \sum Y_{C_v} \left(F_{C_v} - F'_{C_v}\right)^2 + \sum Y_w (F_W - F'_W)^2 + \sum Y_\sigma (F_\sigma - F'_\sigma)^2 + \sum Y_B (F_B - F'_B)^2 \tag{2.11}$$

where Y_p, Y_{Cv}, Y_W, Y_σ, and Y_B are weights applied to each data point. The subscripts P, C_v, W, σ, and B refer to weights for P-ρ-T data, C_v-ρ-T data, W-P-T data, Maxwell criterion data, and virial data, respectively. Often equations of state are fitted to the data with a fixed functional form with specified values for density and temperature exponents. However, more accurate equations of state are possible if these exponents are determined in a selection process. Schmidt and Wagner [2], de Reuck and Armstrong [4], and Bjornn [5] have developed stepwise regression algorithms for this purpose.

2.2.3 Non-linear Least-Squares Regression

Nonlinear fitting is currently the most effective method used to develop equations of state [6]. The techniques used continue to expand as new information is learned about fluid properties at extremely low temperatures and/or high pressures, and as the limits for the temperature exponents and Gaussian bell-shaped terms are refined. The method requires an initial starting point, which is often taken from recently developed equations for another molecularly similar fluid that demonstrates favorable behavior.

Nonlinear fitting does not require the need to linearize experimental data as described in the previous section. Experimental data can be used directly, instead of, for example, requiring a preliminary equation of state to transform pressure and temperature measured experimentally to density and temperature as required by the equation of state. The most important advantage in nonlinear fitting is the ability to use "greater than" or "less than" operators in constraints that control the extrapolation behavior of properties to regions outside that of the range of validity (usually determined as the range of experimental data). Linear fitting can only use equality conditions, and areas without data but with incorrect behavior are often extrapolated graphically to obtain the proper shape. Nonlinear fitting controls the behavior by coercing calculated values to be greater or less than those for lower or higher temperatures (or densities).

Nonlinear algorithms reduce the overall sum of squares of the deviations by adjusting the parameters of the equation of state with a residual sum of squares expression given as

$$S = \sum W_\rho F_\rho^2 + \sum W_p F_p^2 + \sum W_{c_v} F_{c_v}^2 + \cdots \tag{2.12}$$

where W is the weight assigned to each data point and F is the function used to minimize the deviations. The deviations from all data types are summed and weighted such that each property contributes according to its uncertainty.

Because nonlinear methods can fit the temperature exponents and Gaussian-bell-shaped parameters simultaneously with the coefficients of the equations of state, the number of terms that are required to both fit the experimental data and to obtain correct extrapolation is generally half that required in linear fitting. This results in faster calculations and less inter-correlation between the terms.

Constraints form a very essential part of nonlinear fitting equations of state. These have the ability to smooth or change the slope and curvature of any constant property line on the surface of state, as long as such changes do not conflict with behavior caused by fitting accurate experimental data or by other valid constraints. When experimental data are available for derived thermodynamic properties, such as speed of sound and isobaric or isochoric heat capacities, there may be conflicts that must be avoided in the characteristics of parts of the surface, particularly the low temperature liquid or low density states approaching atmospheric pressure. Each of these properties is a function of derivatives of the Helmholtz energy with respect to temperature and/or density, and thus each property requires that the others have the correct characteristic behavior.

On the other hand, fitting the slope and curvature of the speed of sound and other properties below the triple point is possible, and generally required, and the person fitting the equation must be aware of how such properties should behave. Unrealistic shapes far from the region of validity still impact calculated values within the valid region because second order derivatives are part of heat capacity or speed of sound equations. To eliminate small bumps in the surface, non-linear fitting also allows the ability to control the third and fourth derivatives of the equation so that very smooth behavior can be obtained at all conditions outside and inside the range of validity.

2.2.4 Criteria for Equations of State

The quality of a thermodynamic property formulation is determined by its ability to model the physical behavior of the fluid as represented by the available data and its conformance to theory. Published correlations should include estimates of the accuracy of calculated properties as well as a careful definition of the range of validity. A modern thermodynamic property formulation is generally capable of representing all data values within the estimated experimental uncertainty of the measurements (see Table 1.2). The criteria listed below are generally used by correlators in developing equations of state for pure fluids:

1. Pv approaches RT in the limit as ρ approaches zero (i.e., the ideal-gas limit is approached).
2. At low densities the equation of state degenerates to a truncated form of the virial equation.
3. At the critical point
   (a) $(\partial P/\partial v)_T = 0$
   (b) $(\partial^2 P/\partial v^2)_T = 0$
   (c) $(\partial P/\partial T)_{\rho=\rho_c} = dP/dT$ of the vapor pressure curve
4. At the coexistence boundary, the Maxwell criteria (equal Gibbs energies in the liquid and vapor phases at the saturation temperature and pressure) is applied to the equation of state, i.e., coexistence property values predicted by the equation of state should be consistent with those from the ancillary equations for the vapor pressure and the liquid and vapor saturated densities.
5. Numerical requirements (constraints) may be imposed in fitting the equation of state for state points known to high accuracy.

2.2.5 Critical Region Behavior

Correct behavior of the equation of state in the critical region is of concern to some users of property formulations. Because of the nonanalytic behavior of certain fluid properties such as C_v in the critical region, special correlations (e.g., the scaled equations of state presented by Tang et al. [7], Sengers [8]) that use models that account for the fluid behavior at the critical point have been developed for critical region properties. Some provisions can be added to wide range equations of state to approximate critical-region behavior. Wide range models may be improved by including calculated properties near the critical point in the data set used for fitting. Special empirical terms may also be included in equations of state to improve the representation of calculated fluid properties near the critical point. The following statements summarize the considerations used in correlating thermodynamic data in the critical region:

1. The critical point is a singular point (e.g., $C_v \to \infty$, $C_p \to \infty$, $W \to 0$). However, an analytic equation of state cannot fully represent the critical point behavior. Non-analytic terms are required in an equation of state that correctly models the critical region behavior.
2. Most equations of state are analytic and properties may be calculated with differential or integral equations. A non-analytic equation requires numerical solutions for derived property calculations.

3. Critical point phenomena are reflected in the derived properties. The critical region anomalous behavior observed in C_p, C_v, and W generally distort the property surfaces as defined by an analytic equation of state.

2.3 Use of Fundamental Equations of State for Cryogenic Fluids

The equation of state described in this chapter is a general form used for the calculation of properties of a large number of fluids. Although the original references reporting the equations of state for cryogenic fluids use a wide variety of formats, the equations have been transformed to the fundamental form given in Sect. 2.1. Details of the implementation of this equation as a generalized form for cryogenic fluids follow.

Table 2.2 gives the thermodynamic property relations for the fundamental equation summarized in the previous section. The functions in this table are readily used in computer programs for the calculation of thermodynamic properties as given by Lemmon et al. [9, 10]. Equations of state for each of the cryogenic fluids discussed in subsequent chapters have been transformed, if necessary, to this format for consistency.

2.4 Ancillary Functions

In the development of a thermodynamic property formulation for a pure fluid or fluid mixture, it is desirable to determine separate ancillary functions for the liquid-vapor coexistence properties and for the ideal-gas heat capacity. There are many functional forms and methods for determining these ancillary functions. Equations for the vapor pressure, saturated liquid density, and saturated vapor density are used to define the saturation (coexistence) states during the development of the equation of state and as initial estimates in the iterative calculation of saturation properties used in computer programs for the completed property formulation. Equations for the freezing liquid line (also known as the melting line) and sublimation line are not provided here. If calculations exceed the ranges of tabulated properties, the user is encouraged to consult REFPROP or the original publication.

The ancillary functions for vapor pressure, saturated liquid density, and saturated vapor density are given as estimating functions for those properties. These functions are also useful in computer programs for calculating properties iteratively with an equation of state with input properties other than the independent variables of the equation. The default form for most equations is presented here, with the exceptions to these default formats described below and in the fluid specific chapters.

Table 2.2 Functions for calculation of thermodynamic properties

Property	Relation
Compressibility factor	$Z = \frac{P}{\rho RT} = 1 + \delta\left(\frac{\partial \alpha^r}{\partial \delta}\right)$
Pressure	$\frac{P}{P_c} = \frac{\delta}{\tau Z_c}\left[1 + \delta\left(\frac{\partial \alpha^r}{\partial \delta}\right)\right]$
Fugacity coefficient	$\phi = \exp[Z - 1 - \ell n(Z) + \alpha^r]$
Second virial coefficient	$B(T) = \frac{1}{\rho_c}\left(\frac{\partial \alpha^r}{\partial \delta}\right)_{\delta=0}$
Third virial coefficient	$C(T) = \frac{1}{\rho_c^2}\left(\frac{\partial^2 \alpha^r}{\partial \delta^2}\right)_{\delta=0}$
Internal energy	$\frac{U}{RT} = \tau\left[\left(\frac{\partial \alpha^o}{\partial \tau}\right) + \left(\frac{\partial \alpha^r}{\partial \tau}\right)\right]$
Enthalpy	$\frac{H}{RT} = \tau\left[\left(\frac{\partial \alpha^o}{\partial \tau}\right) + \left(\frac{\partial \alpha^r}{\partial \tau}\right)\right] + \delta\left(\frac{\partial \alpha^r}{\partial \delta}\right) + 1$
Gibbs energy	$\frac{G}{RT} = 1 + \alpha^o + \alpha^r + \delta\left(\frac{\partial \alpha^r}{\partial \delta}\right)$
Helmholtz energy	$\frac{a}{RT} = \alpha^0 + \alpha^r$
Entropy	$\frac{S}{R} = \tau\left[\left(\frac{\partial \alpha^o}{\partial \tau}\right) + \left(\frac{\partial \alpha^r}{\partial \tau}\right)\right] - \alpha^o - \alpha^r$
Speed of sound	$\frac{W^2}{RT} = \frac{C_p}{C_v}\left[1 + 2\delta\left(\frac{\partial \alpha^r}{\partial \delta}\right) + \delta^2\left(\frac{\partial^2 \alpha^r}{\partial \delta^2}\right)\right]$
Isochoric heat capacity	$\frac{C_v}{R} = -\tau^2\left[\left(\frac{\partial^2 \alpha^o}{\partial \tau^2}\right) + \left(\frac{\partial^2 \alpha^r}{\partial \tau^2}\right)\right]$
Isobaric heat capacity	$\frac{C_p}{R} = \frac{C_v}{R} + \frac{\left[1 + \delta\left(\frac{\partial \alpha^r}{\partial \delta}\right) - \delta\tau\left(\frac{\partial^2 \alpha^r}{\partial \delta \partial \tau}\right)\right]^2}{\left[1 + 2\delta\left(\frac{\partial \alpha^r}{\partial \delta}\right) + \delta^2\left(\frac{\partial^2 \alpha^r}{\partial \delta^2}\right)\right]}$
First pressure derivative with respect to density	$\left(\frac{\partial P}{\partial \rho}\right)_T = RT\left[1 + 2\delta\left(\frac{\partial \alpha^r}{\partial \delta}\right) + \delta^2\left(\frac{\partial^2 \alpha^r}{\partial \delta^2}\right)\right]$
Second pressure derivative with respect to density	$\left(\frac{\partial^2 P}{\partial \rho^2}\right)_T = \frac{RT}{\rho}\left[2\delta\left(\frac{\partial \alpha^r}{\partial \delta}\right) + 4\delta^2\left(\frac{\partial^2 \alpha^r}{\partial \delta^2}\right) + \delta^3\left(\frac{\partial^3 \alpha^r}{\partial \delta^3}\right)\right]$
First pressure derivative with respect to temperature	$\left(\frac{\partial P}{\partial T}\right)_\rho = R\left[1 + \delta\left(\frac{\partial \alpha^r}{\partial \delta}\right) - \delta\tau\left(\frac{\partial^2 \alpha^r}{\partial \delta \partial \tau}\right)\right]$
Joule-Thomson coefficient	$\mu_j = \left(\frac{\partial T}{\partial P}\right)_h = \frac{T\beta - 1}{\rho C_p}$
Isentropic expansion coefficient	$k = -\frac{v}{P}\left(\frac{\partial P}{\partial v}\right)_s = \frac{W^2 \rho}{P}$
Isothermal expansion coefficient	$k_T = -\frac{v}{P}\left(\frac{\partial P}{\partial v}\right)_T = \frac{\rho}{P}\left(\frac{\partial P}{\partial \rho}\right)_T$
Volume expansivity	$\beta = \frac{1}{v}\left(\frac{\partial v}{\partial T}\right)_P = \frac{1}{\rho}\left(\frac{\partial P}{\partial T}\right)_\rho\left(\frac{\partial P}{\partial \rho}\right)_T$
Adiabatic compressibility	$\beta_s = \frac{1}{kP} = -\frac{1}{v}\left(\frac{\partial v}{\partial P}\right)_s$
Adiabatic bulk modulus	$B_s = kP = -v\left(\frac{\partial P}{\partial v}\right)_s$
Isothermal compressibility	$\kappa = \frac{1}{k_T P} = -\frac{1}{v}\left(\frac{\partial v}{\partial P}\right)_T$
Isothermal bulk modulus	$K_T = k_T P = -v\left(\frac{\partial P}{\partial v}\right)_T$

The generic form of the vapor pressure ancillary function is

$$\ln\left(\frac{p_\sigma}{p_c}\right) = \tau \sum_{i=1}^{q} N_i \theta^{k_i} \tag{2.13}$$

where $\theta = (1\text{–}T/T_c)$, p_σ is the saturated vapor pressure, and the values for the coefficients N_i and the exponents k_i are given in each fluid chapter.

The generic form of the saturated liquid density ancillary function is

$$\frac{\rho'}{\rho_c} = 1 + \sum_{i=1}^{q} N_i \theta^{k_i} \tag{2.14}$$

where ρ' is the saturated liquid density. The ancillary equations for the saturated liquid density for argon, methane, and nitrogen use a different form, as given by

$$\ln\left(\frac{\rho'}{\rho_c}\right) = \sum_{i=1}^{q} N_i \theta^{k_i} \tag{2.15}$$

The generic form of the saturated vapor density ancillary function is

$$\ln\left(\frac{\rho''}{\rho_c}\right) = \sum_{i=1}^{q} N_i \theta^{k_i} \tag{2.16}$$

where ρ'' is the saturated vapor density. The ancillary equations for the saturated vapor density for argon and nitrogen use a different form, as given by

$$\ln\left(\frac{\rho''}{\rho_c}\right) = \tau \sum_{i=1}^{q} N_i \theta^{k_i} \tag{2.17}$$

The default form of the ideal-gas reduced Helmholtz ancillary function was given previously in Eq. (2.9).

References

1. E.W. Lemmon, R. Span, Multi-parameter equations of state for pure fluids and mixtures. Appl. Thermodyn. Fluids, 394–432 (2010)
2. R. Schmidt, W. Wagner, A new form of the equation of state for pure substances and its applications to oxygen. Fluid Phase Equilib. **19**, 175 (1985)
3. J.G. Hust, R.D. McCarty, Curve-fitting techniques and applications to thermodynamics. Cryogenics **7**(4), 200 (1967)
4. K.M. De Reuck, B. Armstrong, A method of correlation using a search procedure, based on a step-wise least-squares technique, and its application to an equation of state for propylene. Cryogenics **19**(9), 505–512 (1979)
5. K.R. Bjornn, A linear least-squares regression algorithm for the optimization of thermodynamic equations of state, M.S. Thesis, University of Idaho, Moscow, Idaho (1988)
6. E.W. Lemmon, M. L. Huber, M.O. McLinden, *Standard Reference Data Program #23, REFPROP Version 9.1* (National Institute of Standards and Technology, Gaithersburg, Maryland 2016)

7. S. Tang, G.X. Jin, J.V. Sengers, Thermodynamic properties of 1,1,1,2-tetrafluoroethane (R134a) in the critical region. Int. J. Thermophys. **12**(3), 515–540 (1991)
8. J.V. Sengers, in *Supercritical Fluids*, ed. by E. Kiran, J.M.H. Levelt Sengers (Kluwer, Dordrecht, 1994), pp. 231–271
9. E.W. Lemmon, R.T. Jacobsen, S.G. Penoncello, S.W. Beyerlein, Computer programs for the calculation of thermodynamic properties of cryogens and other fluids, Adv. Cryo. Eng., **39**:1891–1897 (1994)
10. E.W. Lemmon, R.T Jacobsen, S.G. Penoncello, S.W. Beyerlein, Computer programs for calculating thermodynamic properties of fluids of engineering interest, version 4.1, CATS Report 95-1 (University of Idaho, Moscow, Idaho, 1995)

Chapter 3
Helium

In this work we have chosen helium to denote the classical fluid behavior of the most common isotope ^{4}He. Although the stable isotope ^{3}He is important for radioisotope detectors and fusion energy research, the isotope is so rare that traditional thermophysical property correlation has generally not been necessary. The cost of ^{3}He has varied between \$100 and \$2000 per standard liter and is unlikely to allow bulk characterization research [1]. Traditional statistical thermodynamic approaches are likely to be sufficient for the majority of ^{3}He modeling needs. A preliminary equation of state for ^{3}He has been developed at NIST, but is generally not distributed due to the lack of experimental data available to fully fit the fluid properties.

A unique characteristic of the helium surface of state is the existence of a superfluid liquid phase, known as helium-II, to differentiate from normal a.k.a. classical fluid helium commonly denoted as helium-I. The superfluid properties of helium-II are exceptional and detailed in a companion text of this monograph series [2]. Figure 3.1 shows the phase diagram for helium. The two phases, helium-II and helium-I, are separated by what is known as the lambda (λ) line simply due to similarity of shape with this Greek letter. The λ line is a second order phase transition with a discontinuous temperature dependence on entropy. The intersection of the λ line with the saturation line (2.1768 K) is known as the lower λ point. The lower λ point is often used, out of convenience, as a lower limit to thermodynamic functions, similar to the use of a triple-point temperature as with other fluid functions. However, this characteristic point is not a true triple point as no solid phase is present until 2974 kPa, called the upper λ point. The small region of the helium liquid phase near the upper λ point but below the lower λ point temperature is omitted from the model due to the extreme fluid changes that were not possible to fit during the equation of state development.

J.W. Leachman et al., *Thermodynamic Properties of Cryogenic Fluids*,
International Cryogenics Monograph Series, DOI 10.1007/978-3-319-57835-4_3

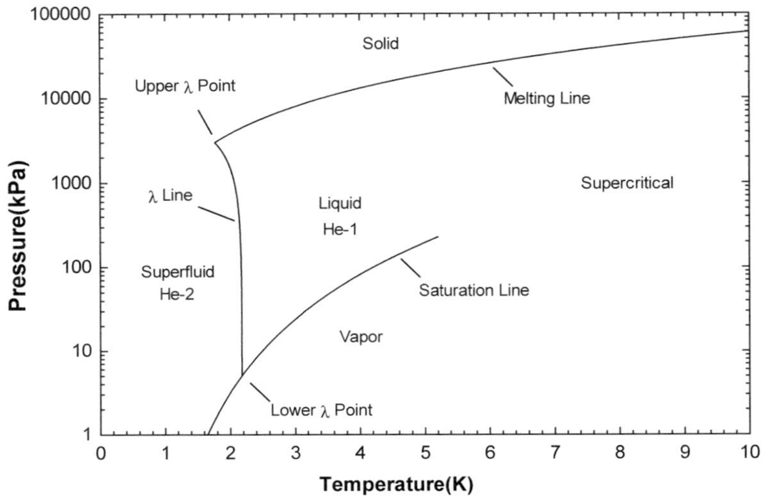

Fig. 3.1 Regions of the helium surface of state at cryogenic temperatures

3.1 Formulation for Helium

A wide-range fundamental equation of state for normal helium was recently developed and is pending journal publication [3]. The uncertainty statement for this formulation in REFPROP follows.

Below 50 K, the uncertainties in density are 0.25% at pressures up to 10 MPa. From 50 to 200 K, the uncertainties decrease linearly from 0.2% at 50 K to 0.05% at 200 K for all pressures up to 50 MPa. Between 200 and 500 K, the uncertainty is 0.03% up to pressures of 40 MPa and 0.1% between 40 and 100 MPa. For all other states not listed here, the uncertainties increase to 0.5% in density. The uncertainties in the speed of sound are 0.01% for the vapor phase and 0.2% for the liquid phase. The uncertainty in vapor pressure is less than 0.02%, and that for heat capacities is about 2%. Uncertainties in the critical region are higher for all properties except vapor pressure (Tables 3.1, 3.2, 3.3 and 3.4).

Table 3.1 Fixed point properties and correlation limits for helium

	Temperature (K)	Pressure (kPa)	Density (mol/dm^3)
Critical point	5.1953	228.32	17.3837
Lower λ point	2.1768	5.0393	36.48
Upper limit	2000	1,000,000	141.22
Lower limit	2.1768		

Table 3.2 Fluid constants and normal boiling point for helium

Molar mass	4.002602 g/mol
Ideal-gas constant	8.3144621 J/(mol K)
Pitzer's acentric factor	−0.3836
Normal boiling point	4.2238 K

Table 3.3 Fundamental equation of state coefficients and exponents for helium

i	N_i	t_i	d_i	p_i	φ_i	β_i	γ_i	D_i
1	0.015559018	1.0	4					
2	3.0638932	0.425	1					
3	−4.2420844	0.63	1					
4	0.054418088	0.69	2					
5	−0.18971904	1.83	2					
6	0.087856262	0.575	3					
7	2.2833566	0.925	1	1				
8	−0.53331595	1.585	1	2				
9	−0.53296502	1.69	3	2				
10	0.99444915	1.51	2	1				
11	−0.30078896	2.9	2	2				
12	−1.6432563	0.8	1	1				
13	0.8029102	1.26	2		1.5497	0.2471	3.15	0.596
14	0.026838669	3.51	1		9.245	0.0983	2.54505	0.3423
15	0.04687678	2.785	2		4.76323	0.1556	1.2513	0.761
16	−0.14832766	1.0	1		6.3826	2.6782	1.9416	0.9747
17	0.03016211	4.22	1		8.7023	2.7077	0.5984	0.5868
18	−0.019986041	0.83	3		0.255	0.6621	2.2282	0.5627
19	0.14283514	1.575	2		0.3523	0.1775	1.606	2.5346
20	0.007418269	3.447	2		0.1492	0.4821	3.815	3.6763
21	−0.22989793	0.73	3		0.05	0.3069	1.61958	4.5245
22	0.79224829	1.634	2		0.1668	0.1758	0.6407	5.039
23	−0.049386338	6.13	2		42.2358	1357.6577	1.076	0.959

Table 3.4 Ancillary function coefficients and exponents for helium

α^0	
i	a_i
0	2.5
1	0.1733486422
2	0.4674523638
$m = 0,\ n = 0$	

	Vapor pressure		Saturated liquid density		Saturated vapor density	
i	N_i	k_i	N_i	k_i	N_i	k_i
1	−4.0558	1.0	2.6970	0.44	−1.8975	−0.41
2	0.69089	1.5	−4.0115	0.72	−6.9133	1.35
3	1.1982	2.0	5.5835	1.0	13.711	2.0
4	0.79703	5.0	−4.7467	1.5	−14.644	2.5
5	−9.0432	14.0	2.7509	2.5	−23.179	10.0
			−2.4711	5.0		

3.2 Property Plots for Helium

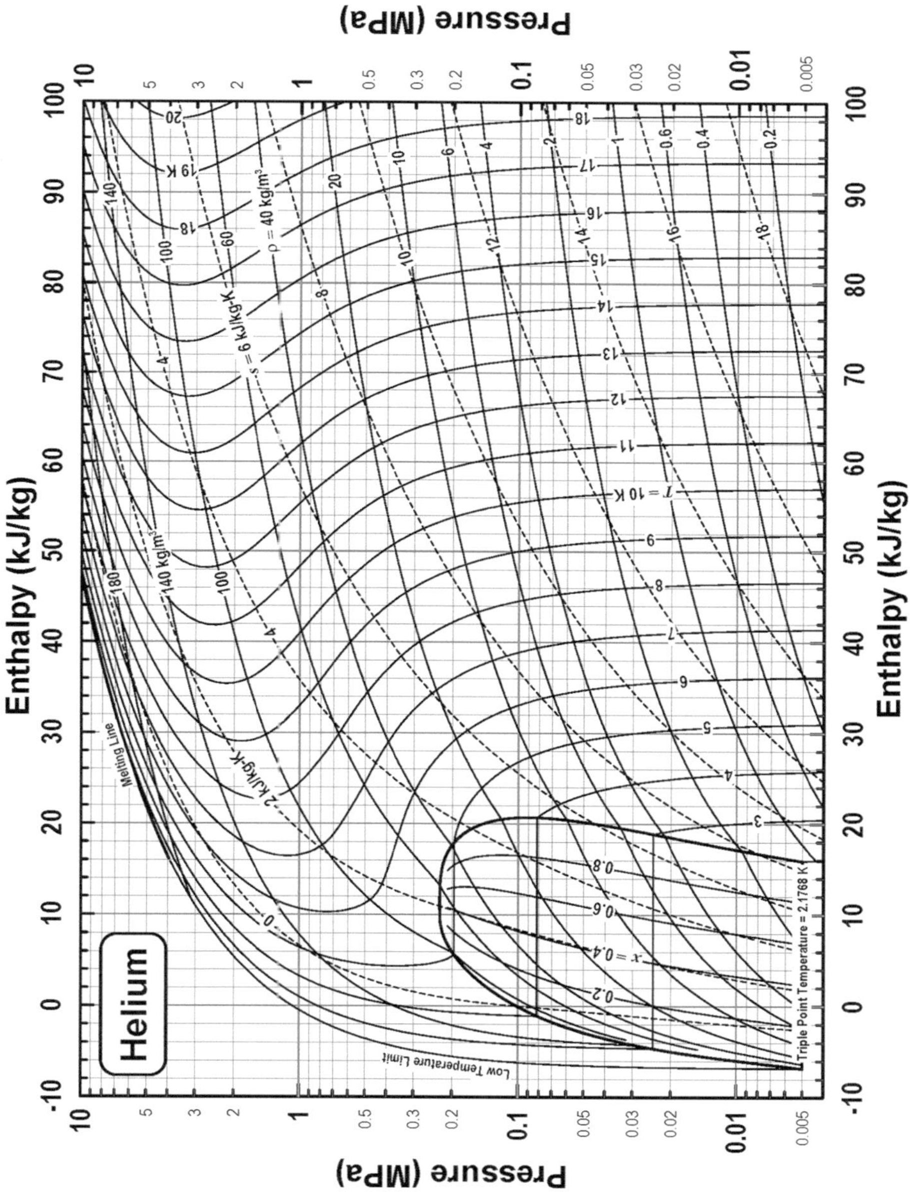
Helium
Enthalpy (kJ/kg)
Pressure (MPa)
Melting Line
Low Temperature Limit
Triple Point Temperature = 2.1768 K
T = 10 K
s = 6 kJ/kg-K
2 kJ/kg-K
ρ = 40 kg/m³
140 kg/m³
x = 0.4

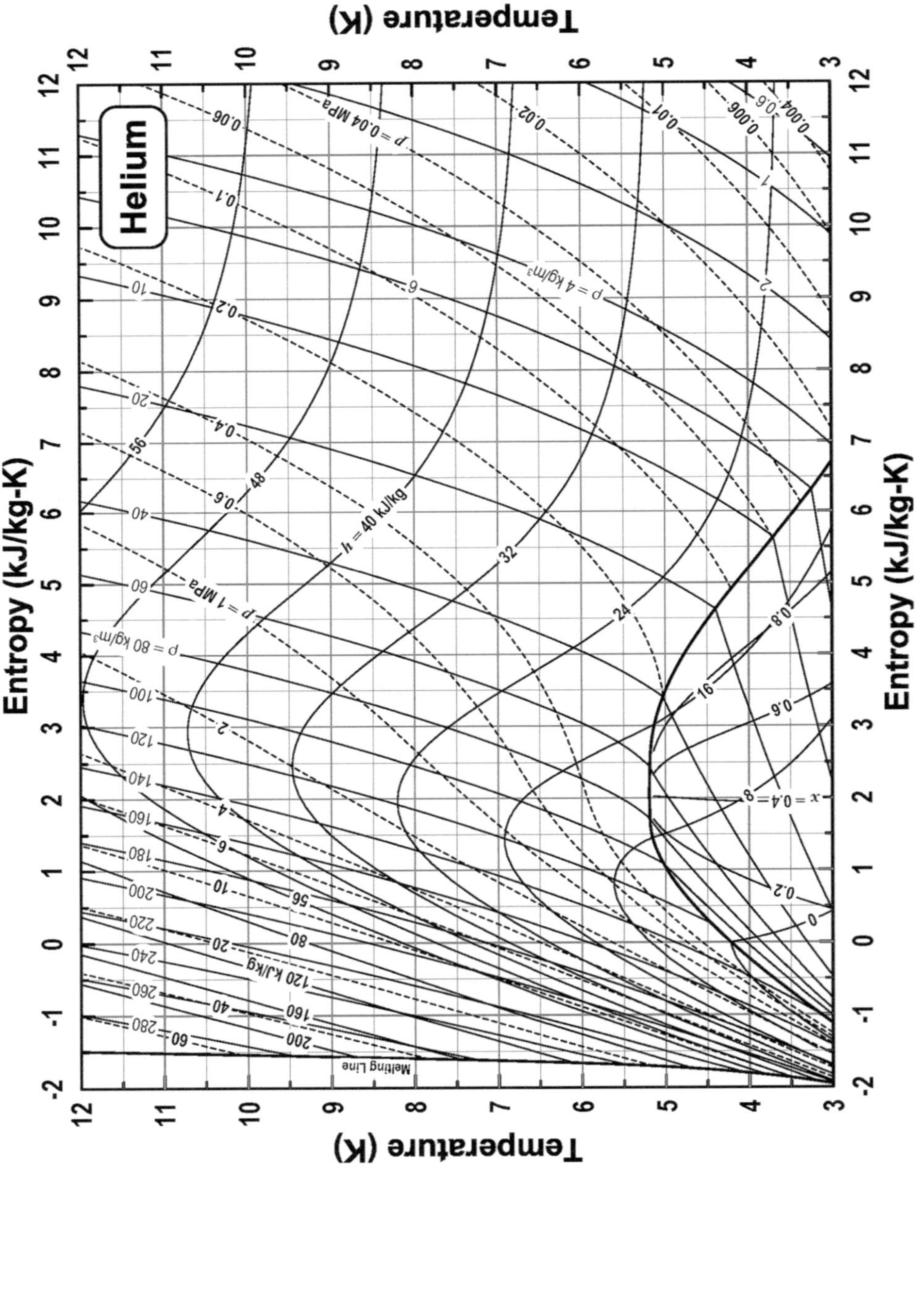
Helium
Entropy (kJ/kg-K)
Temperature (K)
Melting Line
p = 0.04 MPa
p = 1 MPa
ρ = 4 kg/m³
ρ = 80 kg/m³
h = 40 kJ/kg
120 kJ/kg
x = 0.4

3.3 Thermodynamic Data Tables for Helium

See Tables 3.5 and 3.6.

Table 3.5 Thermodynamic properties of saturated helium

Temperature (K)	Pressure (MPa)	Liquid density (kg/m^3)	Vapor density (kg/m^3)	Liquid enthalpy (kJ/kg)	Vapor enthalpy (kJ/kg)	Liquid entropy (kJ/kg K)	Vapor entropy (kJ/kg K)	Liquid C_v (kJ/kg K)	Vapor C_v (kJ/kg K)	Liquid sound speed (m/s)	Vapor sound speed (m/s)
2.1768	0.0050393	146.02	1.1744	−6.8590	15.863	−1.9255	8.5127	3.1487	3.1060	220.55	84.380
2.2	0.0053317	145.99	1.2318	−6.7855	15.960	−1.8929	8.4458	3.0028	3.1059	220.90	84.741
2.4	0.0083507	145.38	1.8014	−6.2368	16.763	−1.6628	7.9204	2.2897	3.1061	223.99	87.672
2.6	0.012376	144.29	2.5160	−5.7427	17.509	−1.4761	7.4667	2.0620	3.1085	225.70	90.283
2.8	0.017562	142.88	3.3936	−5.2387	18.192	−1.3028	7.0653	2.0126	3.1130	224.97	92.589
3.0	0.024061	141.20	4.4535	−4.6994	18.807	−1.1326	6.7028	2.0337	3.1197	222.12	94.601
3.2	0.032024	139.27	5.7188	−4.1119	19.348	−0.96139	6.3698	2.0829	3.1287	217.72	96.326
3.4	0.041599	137.09	7.2176	−3.4673	19.807	−0.78710	6.0583	2.1429	3.1403	212.17	97.769
3.6	0.052936	134.64	8.9863	−2.7576	20.175	−0.60818	5.7620	2.2063	3.1547	205.64	98.927
3.8	0.066186	131.88	11.073	−1.9731	20.439	−0.42306	5.4749	2.2708	3.1722	198.16	99.797
4.0	0.081509	128.74	13.548	−1.1013	20.582	−0.22972	5.1910	2.3361	3.1936	189.64	100.37
4.2	0.099076	125.14	16.511	−0.12442	20.577	−0.025265	4.9035	2.4038	3.2209	179.88	100.63
4.2238	0.10132	124.67	16.902	0.0	20.564	0.0	4.8687	2.4122	3.2246	178.63	100.64
4.4	0.11908	120.91	20.124	0.98481	20.383	0.19482	4.6035	2.4779	3.2574	168.57	100.58
4.6	0.14173	115.80	24.670	2.2723	19.933	0.43828	4.2775	2.5667	3.3100	155.18	100.19
4.8	0.16732	109.26	30.713	3.8270	19.093	0.72054	3.9010	2.6893	3.3894	138.96	99.480
5.0	0.19623	99.841	39.707	5.8799	17.537	1.0827	3.4143	2.9284	3.6197	117.68	97.936
5.1953	0.22832	69.580	69.580	11.543	11.543	2.1157	2.1157				

Table 3.6 Thermodynamic properties of helium

Temperature (K)	Pressure (MPa)	Density (kg/m^3)	Internal energy (kJ/kg)	Enthalpy (kJ/kg)	Entropy (kJ/kg K)	C_v (kJ/kg K)	C_p (kJ/kg K)	Sound speed (m/s)
0.1 MPa								
2.1768	0.1	147.89	−6.9005	−6.2243	−1.9308	2.9490	2.9569	229.68
4.2098	0.1	124.94	−0.8738	−0.07343	−0.0145	2.4073	5.1203	179.37
4.2098	0.1	124.94	−0.8738	−0.07343	−0.0145	2.4073	5.1203	179.37
10	0.1	4.9529	35.167	55.357	10.241	3.1172	5.4122	184.57
20	0.1	2.4111	66.862	108.34	13.919	3.1168	5.2442	263.95
30	0.1	1.6023	98.198	160.61	16.039	3.1171	5.2159	323.40
40	0.1	1.2010	129.45	212.71	17.538	3.1171	5.2059	373.29
50	0.1	0.96081	160.66	264.74	18.699	3.1170	5.2013	417.18
60	0.1	0.80077	191.86	316.74	19.647	3.1169	5.1987	456.83
70	0.1	0.68648	223.05	368.72	20.448	3.1168	5.1972	493.29
80	0.1	0.60076	254.23	420.69	21.142	3.1167	5.1962	527.23
90	0.1	0.53408	285.41	472.64	21.754	3.1167	5.1955	559.11
100	0.1	0.48073	316.58	524.60	22.302	3.1166	5.1950	589.26
110	0.1	0.43707	347.75	576.54	22.797	3.1166	5.1946	617.94
120	0.1	0.40069	378.92	628.49	23.249	3.1165	5.1943	645.34
130	0.1	0.36990	410.09	680.43	23.664	3.1165	5.1941	671.62
140	0.1	0.34350	441.25	732.37	24.049	3.1164	5.1940	696.92
150	0.1	0.32062	472.42	784.31	24.408	3.1164	5.1938	721.33
160	0.1	0.30060	503.58	836.25	24.743	3.1164	5.1937	744.94
170	0.1	0.28293	534.75	888.19	25.058	3.1163	5.1936	767.82
180	0.1	0.26723	565.91	940.12	25.355	3.1163	5.1936	790.04
190	0.1	0.25317	597.07	992.06	25.635	3.1163	5.1935	811.65
200	0.1	0.24053	628.23	1044.0	25.902	3.1163	5.1934	832.70
210	0.1	0.22908	659.40	1095.9	26.155	3.1163	5.1934	853.23
220	0.1	0.21867	690.56	1147.9	26.397	3.1162	5.1934	873.28
230	0.1	0.20917	721.72	1199.8	26.628	3.1162	5.1933	892.88
240	0.1	0.20046	752.88	1251.7	26.849	3.1162	5.1933	912.06
250	0.1	0.19245	784.04	1303.7	27.061	3.1162	5.1933	930.84
260	0.1	0.18505	815.20	1355.6	27.264	3.1162	5.1933	949.25
270	0.1	0.17820	846.36	1407.5	27.460	3.1162	5.1932	967.31
280	0.1	0.17184	877.52	1459.5	27.649	3.1162	5.1932	985.04
290	0.1	0.16592	908.68	1511.4	27.831	3.1162	5.1932	1002.5
300	0.1	0.16039	939.84	1563.3	28.008	3.1161	5.1932	1019.6
400	0.1	0.12031	1251.4	2082.6	29.501	3.1161	5.1931	1177.2
500	0.1	0.096255	1563.0	2601.9	30.660	3.1160	5.1931	1316.0
600	0.1	0.080217	1874.6	3121.3	31.607	3.1160	5.1931	1441.5
700	0.1	0.068760	2186.2	3640.6	32.408	3.1160	5.1931	1557.0
800	0.1	0.060166	2497.8	4159.9	33.101	3.1160	5.1931	1664.5
900	0.1	0.053482	2809.4	4679.2	33.713	3.1160	5.1931	1765.4
1000	0.1	0.048135	3121.0	5198.5	34.260	3.1160	5.1931	1860.9
1200	0.1	0.040113	3744.2	6237.1	35.207	3.1159	5.1931	2038.4
1400	0.1	0.034383	4367.4	7275.7	36.007	3.1159	5.1931	2201.7

(continued)

Table 3.6 (continued)

Temperature (K)	Pressure (MPa)	Density (kg/m^3)	Internal energy (kJ/kg)	Enthalpy (kJ/kg)	Entropy (kJ/kg K)	C_v (kJ/kg K)	C_p (kJ/kg K)	Sound speed (m/s)
1600	0.1	0.030086	4990.5	8314.4	36.701	3.1159	5.1931	2353.7
1800	0.1	0.026743	5613.7	9353.0	37.312	3.1159	5.1931	2496.5
2000	0.1	0.024069	6236.9	10392.	37.859	3.1159	5.1931	2631.5
0.2 MPa								
2.1768	0.2	149.72	−6.9006	−5.5648	−1.9365	2.7684	2.7777	238.47
5.0243	0.2	98.321	4.1582	6.1924	1.1375	2.9872	25.172	114.37
5.0243	0.2	98.321	4.1582	6.1924	1.1375	2.9872	25.172	114.37
10	0.2	10.204	34.080	53.681	8.6916	3.1176	5.6659	183.27
20	0.2	4.8296	66.362	107.77	12.455	3.1177	5.2952	264.83
30	0.2	3.1997	97.876	160.38	14.588	3.1183	5.2383	324.55
40	0.2	2.3971	129.22	212.65	16.092	3.1183	5.2185	374.45
50	0.2	1.9176	160.49	264.79	17.256	3.1181	5.2093	418.30
60	0.2	1.5984	191.73	316.85	18.205	3.1179	5.2042	457.90
70	0.2	1.3705	222.94	368.88	19.007	3.1178	5.2011	494.30
80	0.2	1.1995	254.14	420.88	19.701	3.1176	5.1991	528.18
90	0.2	1.0665	285.34	472.86	20.314	3.1175	5.1977	560.01
100	0.2	0.96011	316.52	524.83	20.861	3.1173	5.1968	590.12
110	0.2	0.87302	347.71	576.80	21.356	3.1172	5.1960	618.75
120	0.2	0.80042	378.88	628.75	21.809	3.1171	5.1955	646.12
130	0.2	0.73897	410.06	680.71	22.224	3.1170	5.1951	672.37
140	0.2	0.68629	441.23	732.66	22.609	3.1170	5.1947	697.64
150	0.2	0.64062	472.40	784.60	22.968	3.1169	5.1945	722.02
160	0.2	0.60065	503.57	836.55	23.303	3.1168	5.1943	745.60
170	0.2	0.56538	534.74	888.49	23.618	3.1168	5.1941	768.46
180	0.2	0.53402	565.91	940.43	23.915	3.1167	5.1939	790.66
190	0.2	0.50596	597.08	992.37	24.196	3.1167	5.1938	812.25
200	0.2	0.48070	628.24	1044.3	24.462	3.1166	5.1937	833.28
210	0.2	0.45784	659.41	1096.2	24.715	3.1166	5.1936	853.80
220	0.2	0.43706	690.57	1148.2	24.957	3.1166	5.1936	873.83
230	0.2	0.41808	721.74	1200.1	25.188	3.1165	5.1935	893.41
240	0.2	0.40068	752.90	1252.0	25.409	3.1165	5.1934	912.58
250	0.2	0.38468	784.06	1304.0	25.621	3.1165	5.1934	931.35
260	0.2	0.36990	815.23	1355.9	25.825	3.1165	5.1934	949.74
270	0.2	0.35621	846.39	1407.8	26.021	3.1164	5.1933	967.79
280	0.2	0.34351	877.55	1459.8	26.209	3.1164	5.1933	985.51
290	0.2	0.33167	908.71	1511.7	26.392	3.1164	5.1933	1002.9
300	0.2	0.32063	939.88	1563.6	26.568	3.1164	5.1932	1020.0
400	0.2	0.24054	1251.5	2083.0	28.062	3.1163	5.1931	1177.5
500	0.2	0.19246	1563.1	2602.3	29.221	3.1162	5.1931	1316.3
600	0.2	0.16040	1874.7	3121.6	30.167	3.1161	5.1930	1441.8
700	0.2	0.13749	2186.3	3640.9	30.968	3.1161	5.1930	1557.2
800	0.2	0.12031	2497.9	4160.2	31.661	3.1161	5.1930	1664.7
900	0.2	0.10695	2809.5	4679.5	32.273	3.1160	5.1931	1765.6

(continued)

Table 3.6 (continued)

Temperature (K)	Pressure (MPa)	Density (kg/m^3)	Internal energy (kJ/kg)	Enthalpy (kJ/kg)	Entropy (kJ/kg K)	C_v (kJ/kg K)	C_p (kJ/kg K)	Sound speed (m/s)
1000	0.2	0.096258	3121.1	5198.8	32.820	3.1160	5.1931	1861.0
1200	0.2	0.080219	3744.2	6237.4	33.767	3.1160	5.1931	2038.6
1400	0.2	0.068761	4367.4	7276.0	34.567	3.1160	5.1931	2201.9
1600	0.2	0.060168	4990.6	8314.6	35.261	3.1160	5.1931	2353.8
1800	0.2	0.053483	5613.8	9353.3	35.873	3.1160	5.1931	2496.6
2000	0.2	0.048136	6236.9	10392.	36.420	3.1159	5.1931	2631.6
0.5 MPa								
2.1768	0.5	154.55	−6.8682	−3.6329	−1.9545	2.3562	2.3698	261.32
10	0.5	28.005	30.494	48.348	6.4314	3.1031	6.6099	181.91
20	0.5	12.115	64.866	106.14	10.479	3.1201	5.4484	267.91
30	0.5	7.9588	96.921	159.74	12.654	3.1218	5.3037	328.16
40	0.5	5.9545	128.53	212.50	14.172	3.1219	5.2551	378.02
50	0.5	4.7638	159.97	264.93	15.342	3.1215	5.2326	421.70
60	0.5	3.9725	191.32	317.19	16.295	3.1210	5.2203	461.11
70	0.5	3.4077	222.62	369.35	17.099	3.1205	5.2127	497.33
80	0.5	2.9840	253.89	421.45	17.795	3.1201	5.2078	531.05
90	0.5	2.6542	285.13	473.51	18.408	3.1198	5.2045	562.73
100	0.5	2.3902	316.36	525.55	18.956	3.1195	5.2021	592.69
110	0.5	2.1741	347.58	577.56	19.452	3.1192	5.2003	621.21
120	0.5	1.9939	378.79	629.55	19.904	3.1190	5.1989	648.47
130	0.5	1.8413	409.99	681.54	20.320	3.1187	5.1979	674.62
140	0.5	1.7104	441.18	733.51	20.706	3.1186	5.1971	699.79
150	0.5	1.5969	472.37	785.48	21.064	3.1184	5.1964	724.09
160	0.5	1.4975	503.56	837.44	21.400	3.1182	5.1959	747.60
170	0.5	1.4098	534.74	889.40	21.714	3.1181	5.1955	770.38
180	0.5	1.3318	565.92	941.35	22.011	3.1180	5.1951	792.52
190	0.5	1.2620	597.10	993.30	22.292	3.1179	5.1948	814.05
200	0.5	1.1991	628.27	1045.2	22.559	3.1178	5.1945	835.02
210	0.5	1.1422	659.45	1097.2	22.812	3.1177	5.1943	855.49
220	0.5	1.0905	690.62	1149.1	23.054	3.1176	5.1941	875.47
230	0.5	1.0432	721.79	1201.1	23.285	3.1175	5.1940	895.01
240	0.5	0.99990	752.96	1253.0	23.506	3.1174	5.1939	914.13
250	0.5	0.96003	784.13	1305.0	23.718	3.1174	5.1937	932.86
260	0.5	0.92321	815.30	1356.9	23.922	3.1173	5.1936	951.22
270	0.5	0.88912	846.47	1408.8	24.118	3.1173	5.1935	969.23
280	0.5	0.85745	877.64	1460.8	24.306	3.1172	5.1935	986.92
290	0.5	0.82797	908.80	1512.7	24.489	3.1172	5.1934	1004.3
300	0.5	0.80044	939.97	1564.6	24.665	3.1171	5.1933	1021.4
400	0.5	0.60072	1251.6	2083.9	26.159	3.1168	5.1930	1178.6
500	0.5	0.48077	1563.2	2603.2	27.317	3.1166	5.1929	1317.3
600	0.5	0.40074	1874.8	3122.5	28.264	3.1164	5.1929	1442.6
700	0.5	0.34355	2186.4	3641.8	29.065	3.1164	5.1929	1558.0
800	0.5	0.30065	2498.0	4161.1	29.758	3.1163	5.1929	1665.3

(continued)

Table 3.6 (continued)

Temperature (K)	Pressure (MPa)	Density (kg/m³)	Internal energy (kJ/kg)	Enthalpy (kJ/kg)	Entropy (kJ/kg K)	C_v (kJ/kg K)	C_p (kJ/kg K)	Sound speed (m/s)
900	0.5	0.26727	2809.6	4680.4	30.370	3.1162	5.1929	1766.2
1000	0.5	0.24056	3121.2	5199.7	30.917	3.1162	5.1929	1861.6
1200	0.5	0.20049	3744.4	6238.3	31.864	3.1161	5.1929	2039.1
1400	0.5	0.17186	4367.6	7276.9	32.664	3.1161	5.1930	2202.3
1600	0.5	0.15039	4990.8	8315.4	33.358	3.1161	5.1930	2354.2
1800	0.5	0.13369	5613.9	9354.0	33.969	3.1160	5.1930	2496.9
2000	0.5	0.12032	6237.1	10393.	34.516	3.1160	5.1930	2631.9
1.0 MPa								
2.1768	1	161.12	−6.7430	−0.53633	−1.9864	1.9281	1.9483	292.34
10	1	61.056	24.470	40.848	4.5031	3.0327	7.5992	198.02
20	1	24.254	62.400	103.63	8.9242	3.1241	5.6901	274.72
30	1	15.764	95.369	158.81	11.165	3.1278	5.4064	334.69
40	1	11.777	127.42	212.33	12.706	3.1279	5.3127	384.18
50	1	9.4262	159.13	265.22	13.886	3.1271	5.2695	427.46
60	1	7.8668	190.67	317.79	14.845	3.1261	5.2458	466.51
70	1	6.7539	222.11	370.17	15.652	3.1252	5.2313	502.41
80	1	5.9183	253.48	422.43	16.350	3.1244	5.2218	535.84
90	1	5.2684	284.80	474.61	16.965	3.1237	5.2153	567.25
100	1	4.7473	316.10	526.74	17.514	3.1230	5.2106	596.99
110	1	4.3204	347.37	578.83	18.010	3.1225	5.2072	625.30
120	1	3.9641	378.62	630.89	18.463	3.1220	5.2045	652.37
130	1	3.6622	409.86	682.92	18.880	3.1216	5.2025	678.36
140	1	3.4031	441.09	734.94	19.265	3.1212	5.2009	703.38
150	1	3.1783	472.31	786.94	19.624	3.1209	5.1996	727.53
160	1	2.9815	503.53	838.93	19.959	3.1206	5.1986	750.91
170	1	2.8076	534.73	890.91	20.275	3.1203	5.1977	773.58
180	1	2.6528	565.93	942.89	20.572	3.1201	5.1970	795.61
190	1	2.5143	597.13	994.85	20.853	3.1198	5.1964	817.04
200	1	2.3895	628.32	1046.8	21.119	3.1196	5.1959	837.92
210	1	2.2766	659.51	1098.8	21.373	3.1195	5.1955	858.30
220	1	2.1738	690.70	1150.7	21.614	3.1193	5.1951	878.20
230	1	2.0799	721.88	1202.7	21.845	3.1191	5.1948	897.66
240	1	1.9938	753.07	1254.6	22.066	3.1190	5.1945	916.71
250	1	1.9145	784.25	1306.6	22.278	3.1189	5.1943	935.37
260	1	1.8413	815.43	1358.5	22.482	3.1187	5.1941	953.67
270	1	1.7735	846.60	1410.4	22.678	3.1186	5.1939	971.62
280	1	1.7105	877.78	1462.4	22.867	3.1185	5.1938	989.25
290	1	1.6519	908.95	1514.3	23.049	3.1184	5.1936	1006.6
300	1	1.5971	940.13	1566.3	23.225	3.1183	5.1935	1023.6
400	1	1.1994	1251.8	2085.6	24.719	3.1177	5.1929	1180.5
500	1	0.96026	1563.5	2604.8	25.878	3.1173	5.1927	1318.8
600	1	0.80063	1875.1	3124.1	26.825	3.1170	5.1926	1444.0
700	1	0.68650	2186.7	3643.4	27.625	3.1168	5.1926	1559.2

(continued)

Table 3.6 (continued)

Temperature (K)	Pressure (MPa)	Density (kg/m³)	Internal energy (kJ/kg)	Enthalpy (kJ/kg)	Entropy (kJ/kg K)	C_v (kJ/kg K)	C_p (kJ/kg K)	Sound speed (m/s)
800	1	0.60084	2498.3	4162.6	28.319	3.1167	5.1926	1666.5
900	1	0.53419	2809.9	4681.9	28.930	3.1166	5.1926	1767.2
1000	1	0.48085	3121.5	5201.1	29.477	3.1165	5.1927	1862.5
1200	1	0.40080	3744.7	6239.7	30.424	3.1164	5.1927	2039.9
1400	1	0.34360	4367.9	7278.2	31.225	3.1163	5.1928	2203.0
1600	1	0.30068	4991.0	8316.8	31.918	3.1162	5.1928	2354.9
1800	1	0.26730	5614.2	9355.3	32.530	3.1162	5.1928	2497.5
2000	1	0.24059	6237.4	10394.	33.077	3.1161	5.1929	2632.5
2.0 MPa								
2.1768	2	171.23	−6.3484	5.3318	−2.0519	1.5123	1.5431	341.10
10	2	108.89	16.983	35.350	2.7846	2.9497	6.5478	264.23
20	2	47.458	57.873	100.01	7.3045	3.1301	5.9852	295.26
30	2	30.758	92.439	157.46	9.6394	3.1406	5.5740	349.76
40	2	22.996	125.32	212.29	11.218	3.1402	5.4129	397.29
50	2	18.441	157.54	265.99	12.416	3.1384	5.3354	439.37
60	2	15.422	189.42	319.11	13.385	3.1363	5.2921	477.50
70	2	13.265	221.11	371.89	14.199	3.1345	5.2655	512.65
80	2	11.644	252.68	424.45	14.900	3.1328	5.2478	545.44
90	2	10.379	284.17	476.86	15.518	3.1314	5.2356	576.32
100	2	9.3643	315.59	529.17	16.069	3.1301	5.2268	605.58
110	2	8.5313	346.97	581.40	16.567	3.1290	5.2202	633.47
120	2	7.8351	378.32	633.58	17.021	3.1281	5.2152	660.16
130	2	7.2444	409.63	685.71	17.438	3.1272	5.2113	685.81
140	2	6.7368	440.93	737.81	17.824	3.1265	5.2082	710.52
150	2	6.2959	472.21	789.88	18.183	3.1258	5.2057	734.40
160	2	5.9093	503.47	841.92	18.519	3.1252	5.2037	757.52
170	2	5.5676	534.73	893.95	18.835	3.1247	5.2020	779.96
180	2	5.2632	565.97	945.96	19.132	3.1242	5.2007	801.77
190	2	4.9905	597.20	997.96	19.413	3.1238	5.1995	823.00
200	2	4.7447	628.43	1050.0	19.680	3.1234	5.1985	843.69
210	2	4.5219	659.65	1101.9	19.933	3.1230	5.1977	863.90
220	2	4.3192	690.86	1153.9	20.175	3.1227	5.1970	883.64
230	2	4.1339	722.07	1205.9	20.406	3.1224	5.1964	902.95
240	2	3.9639	753.28	1257.8	20.627	3.1221	5.1959	921.85
250	2	3.8072	784.48	1309.8	20.839	3.1218	5.1954	940.38
260	2	3.6625	815.68	1361.7	21.043	3.1216	5.1950	958.55
270	2	3.5284	846.87	1413.7	21.239	3.1214	5.1947	976.39
280	2	3.4038	878.06	1465.6	21.428	3.1211	5.1944	993.90
290	2	3.2877	909.25	1517.6	21.610	3.1209	5.1941	1011.1
300	2	3.1792	940.44	1569.5	21.787	3.1208	5.1939	1028.0
400	2	2.3906	1252.2	2088.8	23.281	3.1194	5.1926	1184.1
500	2	1.9155	1563.9	2608.1	24.439	3.1186	5.1921	1321.9
600	2	1.5979	1875.6	3127.3	25.386	3.1181	5.1920	1446.7

(continued)

Table 3.6 (continued)

Temperature (K)	Pressure (MPa)	Density (kg/m³)	Internal energy (kJ/kg)	Enthalpy (kJ/kg)	Entropy (kJ/kg K)	C_v (kJ/kg K)	C_p (kJ/kg K)	Sound speed (m/s)
700	2	1.3706	2187.2	3646.5	26.186	3.1177	5.1920	1561.6
800	2	1.1999	2498.8	4165.7	26.879	3.1175	5.1920	1668.7
900	2	1.0670	2810.4	4684.9	27.491	3.1173	5.1921	1769.2
1000	2	0.96060	3122.0	5204.1	28.038	3.1171	5.1921	1864.4
1200	2	0.80087	3745.2	6242.5	28.985	3.1168	5.1923	2041.5
1400	2	0.68668	4368.4	7281.0	29.785	3.1167	5.1924	2204.5
1600	2	0.60098	4991.6	8319.5	30.478	3.1166	5.1924	2356.2
1800	2	0.53430	5614.8	9358.0	31.090	3.1165	5.1925	2498.7
2000	2	0.48094	6237.9	10396.	31.637	3.1164	5.1926	2633.6
5.0 MPa								
2.4275	5	190.79	−4.4889	21.717	−2.1072	1.1034	1.1607	444.67
10	5	157.72	11.719	43.420	1.3993	2.8516	4.4533	409.39
20	5	100.08	49.019	98.979	5.1876	3.1518	5.9979	375.24
30	5	69.102	85.597	157.95	7.5822	3.1792	5.7686	404.23
40	5	52.884	120.06	214.60	9.2132	3.1783	5.5786	441.36
50	5	42.971	153.42	269.78	10.445	3.1729	5.4651	477.47
60	5	36.262	186.15	324.04	11.434	3.1673	5.3926	511.66
70	5	31.405	218.50	377.71	12.262	3.1623	5.3440	543.94
80	5	27.718	250.58	430.97	12.973	3.1580	5.3101	574.49
90	5	24.819	282.48	483.94	13.597	3.1543	5.2856	603.53
100	5	22.477	314.25	536.70	14.153	3.1512	5.2674	631.25
110	5	20.544	345.92	589.30	14.654	3.1484	5.2536	657.81
120	5	18.920	377.52	641.78	15.111	3.1460	5.2429	683.34
130	5	17.536	409.05	694.17	15.530	3.1439	5.2344	707.95
140	5	16.343	440.53	746.48	15.918	3.1421	5.2276	731.74
150	5	15.302	471.97	798.72	16.278	3.1405	5.2221	754.78
160	5	14.387	503.38	850.92	16.615	3.1390	5.2176	777.14
170	5	13.575	534.76	903.08	16.931	3.1377	5.2138	798.88
180	5	12.851	566.12	955.20	17.229	3.1365	5.2107	820.04
190	5	12.200	597.46	1007.3	17.511	3.1354	5.2081	840.68
200	5	11.612	628.78	1059.4	17.778	3.1344	5.2058	860.83
210	5	11.078	660.09	1111.4	18.032	3.1335	5.2039	880.52
220	5	10.592	691.38	1163.4	18.274	3.1327	5.2023	899.78
230	5	10.146	722.66	1215.5	18.505	3.1319	5.2008	918.65
240	5	9.7365	753.93	1267.5	18.727	3.1312	5.1996	937.13
250	5	9.3588	785.19	1319.5	18.939	3.1306	5.1985	955.27
260	5	9.0092	816.45	1371.4	19.143	3.1300	5.1976	973.07
270	5	8.6849	847.69	1423.4	19.339	3.1294	5.1968	990.55
280	5	8.3831	878.93	1475.4	19.528	3.1289	5.1961	1007.7
290	5	8.1017	910.17	1527.3	19.710	3.1284	5.1954	1024.6
300	5	7.8385	941.40	1579.3	19.886	3.1280	5.1949	1041.3
400	5	5.9165	1253.5	2098.6	21.380	3.1247	5.1917	1195.0
500	5	4.7513	1565.3	2617.7	22.539	3.1227	5.1907	1331.2

(continued)

Table 3.6 (continued)

Temperature (K)	Pressure (MPa)	Density (kg/m³)	Internal energy (kJ/kg)	Enthalpy (kJ/kg)	Entropy (kJ/kg K)	C_v (kJ/kg K)	C_p (kJ/kg K)	Sound speed (m/s)
600	5	3.9694	1877.1	3136.7	23.485	3.1214	5.1904	1454.9
700	5	3.4084	2188.8	3655.8	24.285	3.1205	5.1903	1568.9
800	5	2.9862	2500.5	4174.8	24.978	3.1198	5.1904	1675.3
900	5	2.6571	2812.1	4693.8	25.589	3.1193	5.1905	1775.3
1000	5	2.3933	3123.7	5212.9	26.136	3.1189	5.1906	1870.0
1200	5	1.9967	3746.9	6251.1	27.083	3.1183	5.1909	2046.4
1400	5	1.7128	4370.1	7289.3	27.883	3.1179	5.1911	2208.8
1600	5	1.4996	4993.3	8327.5	28.576	3.1176	5.1914	2360.1
1800	5	1.3336	5616.4	9365.8	29.188	3.1173	5.1915	2502.3
2000	5	1.2006	6239.6	10404.	29.735	3.1171	5.1917	2636.8
10.0 MPa								
3.4675	10	209.75	−0.5357	47.140	−1.9254	1.2691	1.3788	558.76
10	10	189.29	11.263	64.092	0.60767	2.7378	3.6698	542.11
20	10	147.50	43.201	111.00	3.7892	3.1985	5.3914	500.21
30	10	113.69	79.156	167.12	6.0605	3.2430	5.7024	497.52
40	10	91.316	114.42	223.93	7.6952	3.2389	5.6419	516.33
50	10	76.232	148.68	279.86	8.9438	3.2280	5.5469	541.97
60	10	65.487	182.23	334.93	9.9480	3.2170	5.4701	569.18
70	10	57.451	215.27	389.33	10.787	3.2072	5.4131	596.30
80	10	51.206	247.95	443.24	11.507	3.1987	5.3704	622.80
90	10	46.209	280.36	496.77	12.137	3.1915	5.3378	648.56
100	10	42.116	312.58	550.02	12.698	3.1852	5.3124	673.55
110	10	38.698	344.63	603.04	13.204	3.1799	5.2924	697.78
120	10	35.801	376.56	655.88	13.663	3.1752	5.2762	721.31
130	10	33.313	408.39	708.57	14.085	3.1711	5.2631	744.17
140	10	31.151	440.13	761.15	14.475	3.1674	5.2524	766.40
150	10	29.255	471.81	813.63	14.837	3.1642	5.2435	788.05
160	10	27.579	503.43	866.02	15.175	3.1613	5.2360	809.15
170	10	26.086	535.00	918.35	15.492	3.1588	5.2298	829.75
180	10	24.747	566.53	970.62	15.791	3.1564	5.2244	849.87
190	10	23.539	598.02	1022.8	16.073	3.1543	5.2199	869.54
200	10	22.445	629.48	1075.0	16.341	3.1524	5.2160	888.80
210	10	21.448	660.92	1127.2	16.595	3.1506	5.2127	907.66
220	10	20.536	692.33	1179.3	16.838	3.1490	5.2098	926.15
230	10	19.699	723.72	1231.4	17.069	3.1476	5.2072	944.29
240	10	18.927	755.09	1283.4	17.291	3.1462	5.2050	962.11
250	10	18.214	786.44	1335.5	17.503	3.1449	5.2031	979.60
260	10	17.553	817.78	1387.5	17.707	3.1438	5.2014	996.80
270	10	16.938	849.11	1439.5	17.904	3.1427	5.1999	1013.7
280	10	16.365	880.42	1491.5	18.093	3.1416	5.1985	1030.4
290	10	15.829	911.73	1543.5	18.275	3.1407	5.1974	1046.8
300	10	15.328	943.02	1595.4	18.451	3.1398	5.1963	1062.9
400	10	11.639	1255.6	2114.7	19.945	3.1333	5.1904	1212.9

(continued)

Table 3.6 (continued)

Temperature (K)	Pressure (MPa)	Density (kg/m³)	Internal energy (kJ/kg)	Enthalpy (kJ/kg)	Entropy (kJ/kg K)	C_v (kJ/kg K)	C_p (kJ/kg K)	Sound speed (m/s)
500	10	9.3815	1567.7	2633.6	21.103	3.1294	5.1883	1346.6
600	10	7.8566	1879.6	3152.4	22.049	3.1268	5.1877	1468.4
700	10	6.7578	2191.4	3671.2	22.849	3.1250	5.1876	1581.0
800	10	5.9234	2503.1	4190.0	23.542	3.1237	5.1877	1686.2
900	10	5.2801	2814.8	4708.7	24.153	3.1226	5.1879	1785.3
1000	10	4.7595	3126.5	5227.5	24.699	3.1218	5.1882	1879.3
1200	10	3.9753	3749.7	6265.2	25.645	3.1206	5.1887	2054.4
1400	10	3.4128	4372.9	7303.0	26.445	3.1198	5.1892	2215.9
1600	10	2.9897	4996.0	8340.9	27.138	3.1192	5.1896	2366.5
1800	10	2.6598	5619.2	9378.8	27.749	3.1187	5.1899	2508.2
2000	10	2.3955	6242.3	10417.	28.296	3.1184	5.1902	2642.2
50.0 MPa								
8.996	50	273.91	29.504	212.05	−1.4898	2.1093	2.5242	991.28
10	50	271.86	30.771	214.69	−1.2116	2.2789	2.7380	995.03
20	50	251.94	51.699	250.16	1.1906	3.2174	4.1454	993.18
30	50	232.77	80.302	295.11	3.0045	3.4620	4.7755	969.94
40	50	214.80	111.96	344.74	4.4300	3.5208	5.1184	950.33
50	50	198.42	144.97	396.96	5.5944	3.5204	5.3044	939.30
60	50	183.84	178.53	450.51	6.5705	3.5003	5.3930	935.97
70	50	171.00	212.23	504.63	7.4047	3.4748	5.4245	938.22
80	50	159.73	245.88	558.90	8.1294	3.4492	5.4258	944.21
90	50	149.83	279.39	613.10	8.7678	3.4253	5.4128	952.63
100	50	141.09	312.74	667.14	9.3372	3.4038	5.3943	962.63
110	50	133.31	345.92	720.98	9.8504	3.3845	5.3745	973.68
120	50	126.36	378.94	774.63	10.317	3.3672	5.3555	985.44
130	50	120.11	411.81	828.10	10.745	3.3518	5.3378	997.71
140	50	114.46	444.54	881.39	11.140	3.3380	5.3218	1010.4
150	50	109.32	477.15	934.54	11.507	3.3255	5.3073	1023.3
160	50	104.62	509.64	987.54	11.849	3.3143	5.2943	1036.4
170	50	100.32	542.04	1040.4	12.169	3.3040	5.2825	1049.7
180	50	96.365	574.34	1093.2	12.471	3.2947	5.2719	1063.0
190	50	92.710	606.55	1145.9	12.756	3.2862	5.2623	1076.5
200	50	89.325	638.69	1198.4	13.026	3.2784	5.2537	1089.9
210	50	86.180	670.76	1250.9	13.282	3.2712	5.2459	1103.4
220	50	83.251	702.77	1303.4	13.526	3.2646	5.2388	1116.8
230	50	80.516	734.73	1355.7	13.758	3.2584	5.2324	1130.2
240	50	77.956	766.63	1408.0	13.981	3.2527	5.2267	1143.6
250	50	75.554	798.48	1460.3	14.194	3.2474	5.2214	1156.9
260	50	73.297	830.29	1512.4	14.399	3.2425	5.2167	1170.1
270	50	71.172	862.06	1564.6	14.596	3.2379	5.2124	1183.3
280	50	69.166	893.80	1616.7	14.785	3.2335	5.2085	1196.3
290	50	67.271	925.50	1668.8	14.968	3.2295	5.2050	1209.3
300	50	65.477	957.17	1720.8	15.144	3.2257	5.2018	1222.2

(continued)

Table 3.6 (continued)

Temperature (K)	Pressure (MPa)	Density (kg/m^3)	Internal energy (kJ/kg)	Enthalpy (kJ/kg)	Entropy (kJ/kg K)	C_v (kJ/kg K)	C_p (kJ/kg K)	Sound speed (m/s)
400	50	51.693	1272.6	2239.8	16.638	3.1974	5.1818	1346.5
500	50	42.697	1586.5	2757.5	17.793	3.1799	5.1736	1462.5
600	50	36.360	1899.6	3274.7	18.736	3.1681	5.1703	1571.1
700	50	31.654	2212.1	3791.7	19.533	3.1597	5.1693	1673.5
800	50	28.023	2524.4	4308.6	20.223	3.1534	5.1694	1770.5
900	50	25.136	2836.4	4825.6	20.832	3.1486	5.1700	1862.9
1000	50	22.786	3148.3	5342.6	21.377	3.1447	5.1708	1951.2
1200	50	19.192	3771.7	6377.0	22.319	3.1391	5.1727	2117.4
1400	50	16.574	4394.9	7411.7	23.117	3.1351	5.1746	2272.1
1600	50	14.583	5018.0	8446.8	23.808	3.1322	5.1763	2417.3
1800	50	13.017	5641.1	9482.2	24.418	3.1300	5.1778	2554.6
2000	50	11.754	6264.0	10518.	24.963	3.1282	5.1791	2685.0
100.0 MPa								
13.894	100	309.89	60.991	383.68	−1.3815	2.7839	3.6517	1354.1
20	100	298.66	72.682	407.51	0.034275	3.2284	4.1259	1350.2
30	100	283.30	98.384	451.37	1.8065	3.5433	4.5977	1333.7
40	100	269.61	127.91	498.81	3.1693	3.6503	4.8685	1310.8
50	100	256.97	159.29	548.44	4.2761	3.6801	5.0469	1288.4
60	100	245.16	191.67	599.57	5.2079	3.6773	5.1715	1269.7
70	100	234.10	224.59	651.75	6.0120	3.6600	5.2582	1255.6
80	100	223.79	257.80	704.64	6.7182	3.6365	5.3158	1246.1
90	100	214.20	291.14	757.99	7.3465	3.6109	5.3509	1240.8
100	100	205.30	324.51	811.60	7.9114	3.5851	5.3691	1238.8
110	100	197.05	357.86	865.33	8.4235	3.5603	5.3753	1239.5
120	100	189.42	391.14	919.08	8.8912	3.5370	5.3732	1242.4
130	100	182.34	424.34	972.78	9.3210	3.5153	5.3658	1246.8
140	100	175.77	457.44	1026.4	9.7183	3.4951	5.3553	1252.5
150	100	169.66	490.45	1079.9	10.087	3.4766	5.3430	1259.2
160	100	163.97	523.36	1133.2	10.432	3.4594	5.3300	1266.5
170	100	158.66	556.18	1186.5	10.754	3.4437	5.3169	1274.5
180	100	153.69	588.91	1239.6	11.058	3.4291	5.3043	1282.8
190	100	149.03	621.55	1292.6	11.344	3.4156	5.2922	1291.5
200	100	144.65	654.10	1345.4	11.616	3.4031	5.2810	1300.5
210	100	140.53	686.58	1398.2	11.873	3.3915	5.2705	1309.7
220	100	136.64	718.99	1450.8	12.118	3.3808	5.2608	1319.0
230	100	132.97	751.34	1503.4	12.352	3.3707	5.2520	1328.5
240	100	129.49	783.61	1555.9	12.575	3.3613	5.2439	1338.0
250	100	126.19	815.83	1608.3	12.789	3.3526	5.2364	1347.7
260	100	123.06	848.00	1660.6	12.994	3.3443	5.2297	1357.4
270	100	120.08	880.11	1712.9	13.191	3.3366	5.2235	1367.2
280	100	117.25	912.18	1765.1	13.381	3.3294	5.2178	1377.1
290	100	114.54	944.20	1817.2	13.564	3.3225	5.2126	1386.9
300	100	111.96	976.17	1869.3	13.741	3.3160	5.2078	1396.9

(continued)

Table 3.6 (continued)

Temperature (K)	Pressure (MPa)	Density (kg/m³)	Internal energy (kJ/kg)	Enthalpy (kJ/kg)	Entropy (kJ/kg K)	C_v (kJ/kg K)	C_p (kJ/kg K)	Sound speed (m/s)
400	100	91.385	1294.1	2388.4	15.234	3.2672	5.1765	1496.5
500	100	77.191	1609.7	2905.2	16.388	3.2362	5.1620	1594.6
600	100	66.799	1924.0	3421.0	17.328	3.2150	5.1552	1689.7
700	100	58.859	2237.4	3936.4	18.122	3.1996	5.1525	1781.3
800	100	52.594	2550.2	4451.6	18.810	3.1880	5.1518	1869.5
900	100	47.526	2862.7	4966.8	19.417	3.1790	5.1522	1954.6
1000	100	43.343	3174.8	5482.0	19.960	3.1718	5.1532	2036.6
1200	100	36.842	3798.6	6512.9	20.900	3.1610	5.1559	2192.8
1400	100	32.026	4422.0	7544.4	21.695	3.1534	5.1589	2339.7
1600	100	28.317	5045.0	8576.5	22.384	3.1478	5.1617	2478.7
1800	100	25.374	5668.0	9609.1	22.992	3.1435	5.1642	2610.9
2000	100	22.982	6290.8	10642.	23.536	3.1402	5.1664	2737.1

References

1. D.A. Shea, D. Morgan, The Helium-3 shortage: supply, demand, and options for congress, Congressional Research Service, 7–5700, R41419, https://fas.org/sgp/crs/misc/R41419.pdf (2010)
2. S.W. Van Sciver, in *Helium Cryogenics*, 2nd edn. (Springer, Berlin, 2012)
3. D.O. Ortiz-Vega, K.R. Hall, J.C. Holste, V.D. Arp, E.W. Lemmon, Interim equation, final equation of state to be published in J. Phys. Chem. Ref. Data.

Chapter 4
Hydrogen

Hydrogen is the simplest and most abundant element in the known universe. In this Chapter we focus on the diatomic molecular form of hydrogen commonly utilized in cryogenics. The simplicity of hydrogen allows small changes in the molecule to have significant effects on thermodynamic properties. Isotopic influences are discussed in Chap. 5 with Deuterium. However, even the behavior of isotopically pure hydrogen is complicated due to the role of quantum mechanics in molecular energy levels. The role of quantum mechanics on the statistical thermodynamics of cryogenic hydrogen is a rich story in the annals of science [1]. Here we briefly cover the basics to assist correct use of the property formulations.

Hydrogen molecules obey Fermi-Dirac statistics that require the overall wave-function of a molecule to be anti-symmetric. This requirement allows Heisenberg's nuclear spin function parity with the rotational spin function of the molecule resulting in two separable forms of hydrogen. Parahydrogen, from the Greek root for abnormal or uncommon, is a lower energy singlet most prevalent in equilibrium below 77 K and restricted to even rotational energy levels (J = 0, 2, 4…). Orthohydrogen, from the Greek root for straight or right, is a higher energy triplet most prevalent in equilibrium above 77 K and restricted to odd rotational energy levels (J = 1, 3, 5…). Figure 3.1 shows the equilibrium orthohydrogen fraction versus temperature. Insufficient thermal energy near the normal boiling points is available to fully populate all states and the equilibrium composition is 99.8% parahydrogen. When sufficient energy is available to fully populate the accessible states near room temperature, a 3:1 ratio of ortho to parahydrogen exists that is commonly known as normal hydrogen. Although spontaneous nuclear spin exchange is forbidden, homogenous conversion from ortho to parahydrogen occurs slowly over the course of days to weeks, and heterogenous conversion is rapidly achieved through use of a catalyst. This ability to manipulate ortho-parahydrogen conversion allows preparation of samples with properties unique to the ortho-parahydrogen composition.

Figure 4.1 shows the differences in ideal-gas isobaric heat capacities of hydrogen at cryogenic temperatures. Caloric properties, particularly heat capacities and thermal conductivities, can display deviations between ortho and parahydrogen as

J.W. Leachman et al., *Thermodynamic Properties of Cryogenic Fluids*,
International Cryogenics Monograph Series, DOI 10.1007/978-3-319-57835-4_4

large as 30% near 140 K. These differences are due to statistical distributions between rotational energy modes and are directly proportional to composition.

Figure 4.1 contains a line labeled "Equilibrium". Equilibrium hydrogen is defined as a sample comprising the equilibrium ortho-parahydrogen ratio at the sample temperature (labeled Equilibrium Orthohydrogen Fraction in the figure). This presents a point of confusion when utilizing hydrogen property formulations that is illustrated by the following example: A sample of hydrogen near 77 K has a 50:50 ratio of ortho to parahydrogen, the equilibrium fraction at that temperature. Is the ideal-gas heat capacity given by the curve labeled "Equilibrium" or "50–50"? The heat capacity measured within any isothermal calorimeter will be the non-equilibrium, composition-specific curve values and not the Equilibrium curve value. The Equilibrium curve is created from summing all of the energy modes within the same statistical partition function, assuming free exchange between all of the energy modes, instead of segregating the para (even) and orthohydrogen (odd) rotational contributions separately. As stated, free transition between ortho and parahydrogen is forbidden by quantum mechanics due to the large energy change required for conversion. The Equilibrium curve is essentially an effective heat capacity that includes this heat of conversion, which is why the peak in heat capacity coincides with the maximum slope of the equilibrium fraction curve. Realizing the Equilibrium curve would require changing temperature during continuous exposure to a perfect catalyst ensuring total conversion. This is clearly an unrealistic condition that is only possible in ideal liquefier analyses. Although equilibrium ideal-gas contributions to α^0 could be developed for use with equations of state, the peak in heat capacity is sufficiently close to the critical point to cause non-physical property behavior and should be avoided.

Ortho and parahydrogen also display real-fluid property differences. Vapor pressure deviations approach 3% near the critical point and fixed point properties shown in Tables 4.1 and 4.7 also display differences. These deviations must be accounted for by

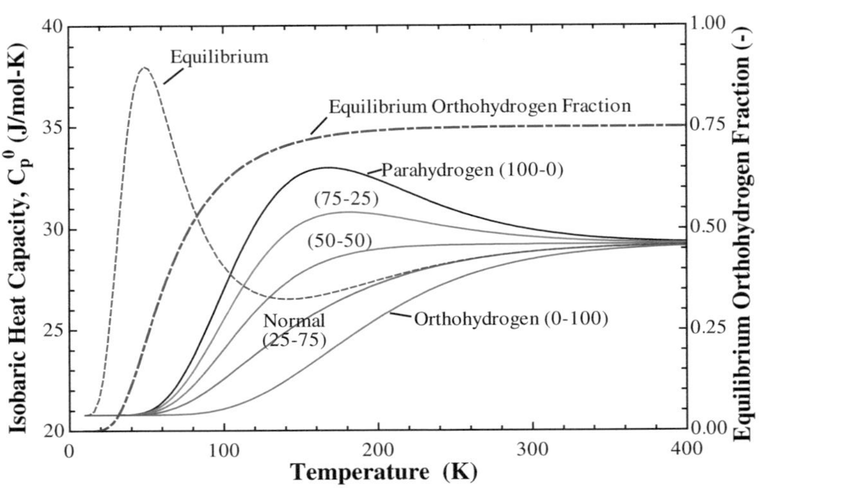

Fig. 4.1 Ideal-gas isobaric heat capacities and equilibrium composition of hydrogen versus temperature

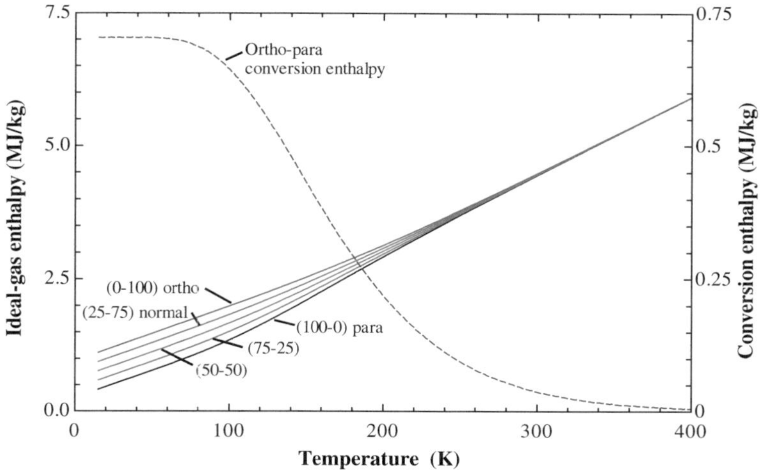

Fig. 4.2 Ideal-gas enthalpies of hydrogen versus temperature

the residual contribution of the equation of state, α^r, and necessitated independent equations of state to be developed for ortho and parahydrogen. As a result, care should be taken when setting reference states of the formulations if comparing or mixing properties. Many property packages, including REFPROP, assume a default reference state of a saturated liquid at the normal boiling point. This assumption will lead to inaccurate differences between ortho and parahydrogen fundamental property predictions where reference states are required. To account for this, the reference states can be user specified under the Options tab in REFPROP. For all comparisons and mixing with parahydrogen, the reference enthalpy and entropy of saturated liquid orthohydrogen at the normal boiling point should be changed to 702.98 kJ/kg and 0.018269 kJ/kg-K, respectively. Figure 4.2 shows the ideal-gas enthalpies and heat of conversion between ortho and parahydrogen at cryogenic temperatures.

4.1 Formulation for Parahydrogen

A 14-term fundamental equation of state was developed for parahydrogen, normal hydrogen, and orthohydrogen by Leachman et al. [2]. The uncertainty statement in REFPROP for this formulation follows.

Table 4.1 Fixed point properties and correlation limits for parahydrogen

	Temperature (K)	Pressure (kPa)	Density (mol/dm^3)
Critical point	32.938	1285.8	15.538
Triple point	13.8033	7.041	
Upper limit	1000	2,000,000	104.0
Lower limit	13.8033		

Table 4.2 Fluid constants and normal boiling point for parahydrogen

Molar mass	2.01588 g/mol
Ideal-gas constant	8.314472 J/(mol K)
Pitzer's acentric factor	−0.219
Normal boiling point	20.271 K

Table 4.3 Fundamental equation of state coefficients and exponents for parahydrogen

i	N_i	t_i	d_i	p_i	φ_i	β_i	γ_i	D_i
1	−7.33375	0.6855	1					
2	0.01	1	4					
3	2.60375	1	1					
4	4.66279	0.489	1					
5	0.682390	0.774	2					
6	−1.47078	1.133	2					
7	0.135801	1.386	3					
8	−1.05327	1.619	1	1				
9	0.328239	1.162	3	1				
10	−0.0577833	3.96	2		1.7437	0.194	0.8048	1.5487
11	0.0449743	5.276	1		0.5516	0.2019	1.5248	0.1785
12	0.0703464	0.99	3		0.0634	0.0301	0.6648	1.28
13	−0.0401766	6.791	1		2.1341	0.2383	0.6832	0.6319
14	0.119510	3.19	1		1.777	0.3253	1.493	1.7104

Table 4.4 Ancillary function coefficients and exponents for parahydrogen

	α^0	
i	a_i	k_i
0	2.5	
1	−1.4485891134	
2	1.884521239	
3	4.30256	499.0
4	13.0289	826.5
5	−47.7365	970.8
6	50.0013	1166.2
7	−18.6261	1341.4
8	0.993973	5395.0
9	0.536078	10185.0

$n = 9$ (there are no polynomial terms in the first summation)

	Vapor pressure		Saturated liquid density		Saturated vapor density	
i	N_i	k_i	N_i	k_i	N_i	k_i
1	−4.87767	1	−0.13509	0.15	−5.7545	0.53
2	1.03359	1.5	4.0739	0.44	3.8153	0.7
3	0.82668	2.65	−5.3985	0.7	−12.293	1.7
4	−0.129412	7.4	5.523	0.99	15.095	2.4
5			−2.3643	1.31	−17.295	3.3
6					−34.19	10

The uncertainty in density is 0.1% at temperatures from the triple point to 250 K and at pressures up to 40 MPa, except in the critical region, where an uncertainty of 0.2% in pressure is generally attained. In the region between 250 and 450 K and at pressures to 300 MPa, the uncertainty in density is 0.04%. At temperatures between 450 and 1000 K, the uncertainty in density increases to 1%. At pressures between 300 and 2000 MPa, the uncertainty in density is 8%. Speed of sound data are represented within 0.5% below 100 MPa. The estimated uncertainty for heat capacities is 1.0%. The estimated uncertainties of vapor pressures and saturated liquid densities calculated with the Maxwell criterion are 0.1% for each property (Tables 4.2, 4.3 and 4.4).

4.2 Property Plots for Parahydrogen

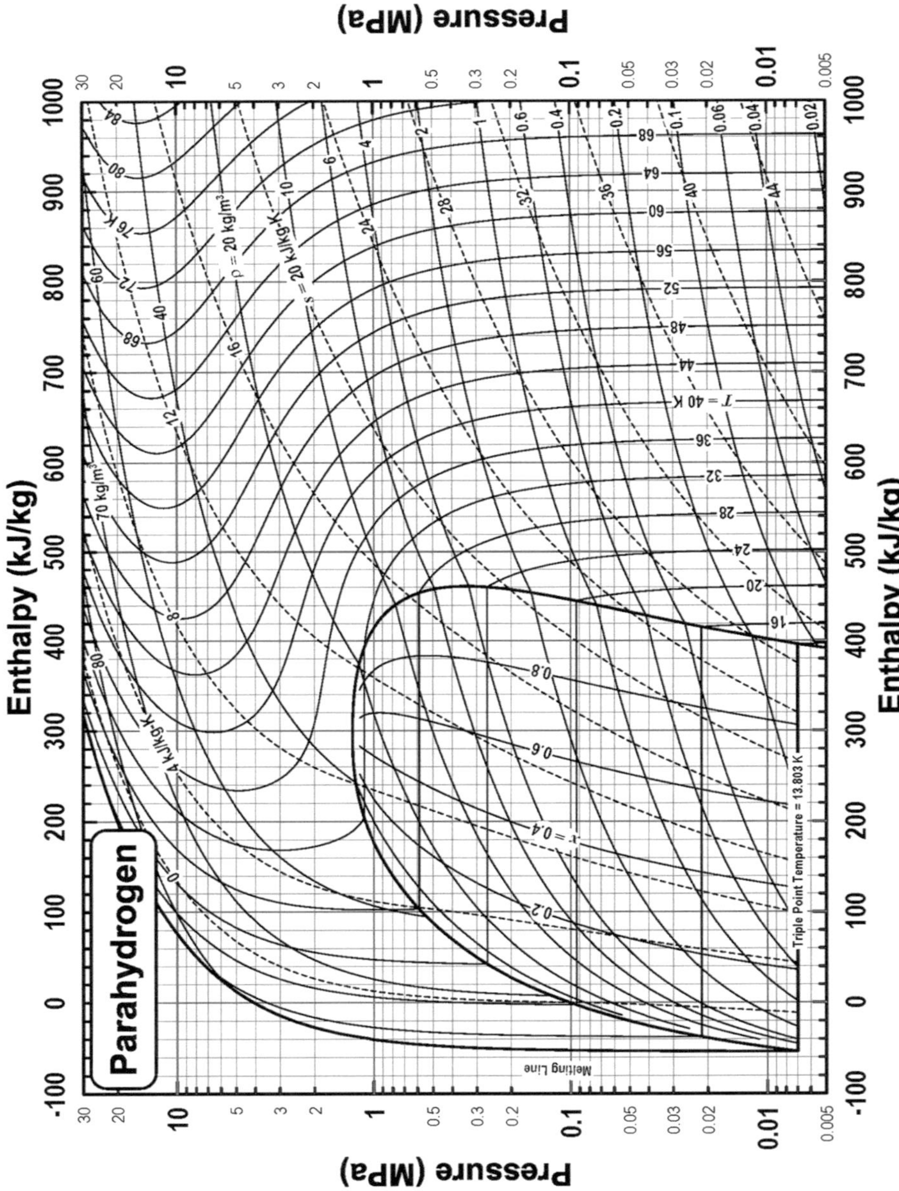
Parahydrogen
Pressure (MPa)
Enthalpy (kJ/kg)
Melting Line
Triple Point Temperature = 13.803 K
T = 40 K
ρ = 20 kg/m3
s = 20 kJ/kg-K
70 kg/m3
4 kJ/kg-K
x = 0.4

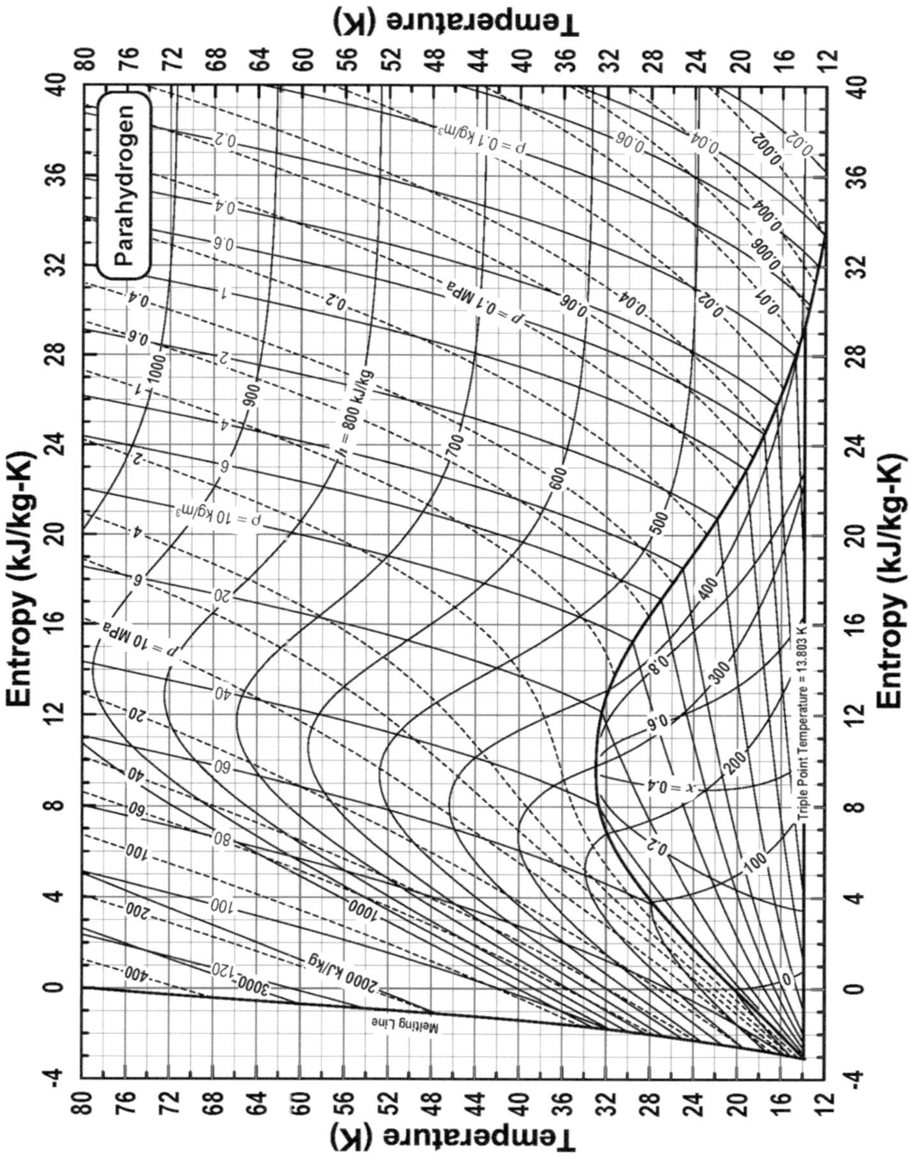
Parahydrogen
Entropy (kJ/kg-K)
Temperature (K)
Triple Point Temperature = 13.803 K
Melting Line

4.3 Thermodynamic Data Tables for Parahydrogen

See Tables 4.5 and 4.6.

Table 4.5 Thermodynamic properties of saturated parahydrogen

Temperature (K)	Pressure (MPa)	Liquid density (kg/m^3)	Vapor density (kg/m^3)	Liquid enthalpy (kJ/kg)	Vapor enthalpy (kJ/kg)	Liquid entropy (kJ/kg K)	Vapor entropy (kJ/kg K)	Liquid C_v (kJ/kg K)	Vapor C_v (kJ/kg K)	Liquid sound speed (m/s)	Vapor sound speed (m/s)
13.803	0.007041	76.977	0.12555	−53.741	396.31	−3.084	29.521	5.1313	6.2265	1263.1	305.65
14	0.007884	76.819	0.13878	−52.364	398.14	−2.9857	29.193	5.1576	6.2358	1256.9	307.56
15	0.013434	75.996	0.2224	−45.167	407.19	−2.4943	27.663	5.2616	6.2818	1231.3	316.84
16	0.021548	75.133	0.33766	−37.603	415.83		26.326	5.3343	6.3225	1210.4	325.43
17	0.032886	74.22	0.49069	−29.626	423.97	−1.5389	25.143	5.4003	6.3569	1190.1	333.39
18	0.048148	73.252	0.688	−21.184	431.52	−1.0684	24.082	5.472	6.387	1168.5	340.72
19	0.068071	72.225	0.93656	−12.224	438.41	−0.59883	23.119	5.5516	6.4167	1144.7	347.43
20	0.093414	71.135	1.244	−2.6915	444.54	−0.12814	22.234	5.6371	6.4499	1118.6	353.5
20.271	0.10133	70.828	1.3386	0	446.07	0	22.005	5.6608	6.4599	1111.1	355.04
21	0.12496	69.977	1.6188	7.4689	449.84	0.34567	21.411	5.7248	6.4902	1090.1	358.94
22	0.1635	68.743	2.0708	18.318	454.2	0.82441	20.637	5.8115	6.5406	1059.2	363.71
23	0.20983	67.425	2.6114	29.925	457.53	1.31	19.902	5.8949	6.6036	1025.9	367.83
24	0.26478	66.01	3.2546	42.371	459.71	1.8045	19.194	5.9739	6.6814	989.95	371.27
25	0.32917	64.485	4.0178	55.761	460.61	2.3107	18.505	6.0485	6.7759	951.23	374.03
26	0.40384	62.827	4.9236	70.23	460.03	2.8321	17.824	6.12	6.8895	909.32	376.1
27	0.48965	61.006	6.0029	85.961	457.73	3.3733	17.143	6.1908	7.0254	863.64	377.48
28	0.5875	58.98	7.2998	103.22	453.35	3.9414	16.446	6.266	7.1879	813.34	378.15
29	0.69833	56.677	8.883	122.41	446.33	4.5475	15.717	6.3541	7.384	757.14	378.08
30	0.82319	53.976	10.871	144.24	435.71	5.2108	14.926	6.4715	7.6246	693.04	377.2
31	0.96329	50.626	13.508	170.13	419.51	5.9713	14.016	6.6543	7.9291	617.45	375.35
32	1.1203	45.901	17.492	204.08	392.26	6.9452	12.826	7.0097	8.3364	522.59	372.03
32.938	1.2858	31.323	31.323	295.63	295.63	9.6253	9.6253				

Table 4.6 Thermodynamic properties of parahydrogen

Temperature (K)	Pressure (MPa)	Density (kg/m^3)	Internal energy (kJ/kg)	Enthalpy (kJ/kg)	Entropy (kJ/kg K)	C_v (kJ/kg K)	C_p (kJ/kg K)	Sound speed (m/s)
0.1 MPa								
13.831	0.1	77.033	−53.816	−52.518	−3.0828	5.1415	6.9236	1264.7
20	0.1	71.144	−4.0348	−2.6292	−0.12965	5.6371	9.5656	1119
20.227	0.1	70.879	−1.8541	−0.4432	−0.02096	5.6569	9.7025	1112.4
20.227	0.1	1.3228	370.22	445.82	22.042	6.4583	12.01	354.79
30	0.1	0.83511	435.3	554.9	26.456	6.213	10.787	447.83
40	0.1	0.61566	499.07	661.49	29.524	6.2184	10.572	521.37
50	0.1	0.48388	562.39	766.94	31.877	6.2841	10.546	583.5
60	0.1	0.40593	626.72	873.07	33.811	6.5031	10.718	635.75
70	0.1	0.34726	694.13	982.1	35.491	6.9414	11.13	678.63
80	0.1	0.3035	766.94	1096.4	37.017	7.5978	11.77	713.92
90	0.1	0.26959	847.09	1218	38.448	8.4083	12.569	744.38
100	0.1	0.24252	935.67	1348	39.817	9.2769	13.43	772.53
110	0.1	0.2204	1032.8	1486.5	41.136	10.108	14.256	799.95
120	0.1	0.20199	1137.7	1632.8	42.408	10.832	14.975	827.42
130	0.1	0.18643	1249.1	1785.5	43.631	11.406	15.546	855.22
140	0.1	0.1731	1365.4	1943.2	44.799	11.819	15.957	883.38
150	0.1	0.16155	1485.1	2104.2	45.909	12.081	16.217	911.82
160	0.1	0.15144	1606.7	2267.1	46.961	12.212	16.346	940.43
170	0.1	0.14253	1729.1	2430.7	47.953	12.238	16.371	969.1
180	0.1	0.13461	1851.3	2594.2	48.888	12.188	16.319	997.7
190	0.1	0.12752	1972.8	2756.9	49.767	12.084	16.215	1026.1
200	0.1	0.12115	2093	2918.4	50.596	11.947	16.077	1054.3
210	0.1	0.11538	2211.7	3078.4	51.376	11.793	15.923	1082.1
220	0.1	0.11013	2328.9	3236.9	52.113	11.634	15.763	1109.5
230	0.1	0.10535	2444.4	3393.7	52.811	11.478	15.607	1136.5
240	0.1	0.10096	2558.5	3549	53.472	11.331	15.459	1162.9
250	0.1	0.09692	2671.1	3702.9	54.1	11.195	15.323	1188.7
260	0.1	0.093193	2782.5	3855.5	54.698	11.073	15.201	1214.1
270	0.1	0.089742	2892.7	4007	55.27	10.965	15.092	1238.8
280	0.1	0.086538	3001.9	4157.4	55.817	10.871	14.997	1263
290	0.1	0.083555	3110.2	4307	56.342	10.789	14.916	1286.7
300	0.1	0.080771	3217.7	4455.8	56.846	10.719	14.845	1309.8
400	0.1	0.060585	4270.5	5921.1	61.064	10.426	14.551	1518.2
500	0.1	0.048472	5311.1	7374.2	64.307	10.401	14.526	1697.7
600	0.1	0.040396	6352.2	8827.7	66.957	10.425	14.549	1859.1
700	0.1	0.034626	7397.1	10285	69.203	10.479	14.604	2006.4
800	0.1	0.030299	8449.3	11750	71.159	10.572	14.697	2142.2
900	0.1	0.026933	9512.9	13226	72.897	10.705	14.829	2268.1
1000	0.1	0.024241	10591	14717	74.468	10.871	14.995	2385.7
0.2 MPa								
13.864	0.2	77.091	−53.772	−51.178	−3.0798	5.1523	6.9238	1266.5
20	0.2	71.279	−4.486	−1.6801	−0.15241	5.6368	9.5182	1124.3
22.802	0.2	67.693	24.608	27.563	1.2132	5.8787	11.475	1032.7

(continued)

Table 4.6 (continued)

Temperature (K)	Pressure (MPa)	Density (kg/m^3)	Internal energy (kJ/kg)	Enthalpy (kJ/kg)	Entropy (kJ/kg K)	C_v (kJ/kg K)	C_p (kJ/kg K)	Sound speed (m/s)
22.802	0.2	2.4967	376.85	456.96	20.045	6.5901	13.408	367.07
30	0.2	1.7353	428.55	543.8	23.366	6.2542	11.382	441.18
40	0.2	1.2513	494.51	654.35	26.55	6.24	10.841	518.55
50	0.2	0.98586	558.97	761.84	28.949	6.2987	10.702	582.31
60	0.2	0.81557	624.01	869.24	30.907	6.5132	10.82	635.4
70	0.2	0.6963	691.9	979.13	32.6	6.9486	11.202	678.73
80	0.2	0.60786	765.05	1094.1	34.134	7.6032	11.823	714.27
90	0.2	0.53955	845.46	1216.1	35.571	8.4126	12.61	744.88
100	0.2	0.48515	934.24	1346.5	36.944	9.2804	13.462	773.13
110	0.2	0.44077	1031.5	1485.3	38.266	10.111	14.282	800.62
120	0.2	0.40388	1136.6	1631.8	39.54	10.834	14.997	828.14
130	0.2	0.37271	1248.1	1784.7	40.764	11.408	15.564	855.98
140	0.2	0.34602	1364.5	1942.5	41.933	11.821	15.972	884.17
150	0.2	0.32291	1484.3	2103.7	43.045	12.083	16.23	912.64
160	0.2	0.3027	1606	2266.7	44.097	12.213	16.357	941.27
170	0.2	0.28488	1728.4	2430.5	45.09	12.24	16.381	969.95
180	0.2	0.26904	1850.7	2594.1	46.025	12.189	16.328	998.56
190	0.2	0.25488	1972.1	2756.8	46.905	12.085	16.222	1027
200	0.2	0.24213	2092.4	2918.4	47.734	11.948	16.084	1055.2
210	0.2	0.2306	2211.2	3078.5	48.515	11.794	15.929	1083
220	0.2	0.22012	2328.4	3237	49.252	11.635	15.769	1110.4
230	0.2	0.21055	2444	3393.9	49.95	11.479	15.612	1137.3
240	0.2	0.20178	2558.1	3549.2	50.611	11.332	15.464	1163.7
250	0.2	0.19372	2670.7	3703.2	51.239	11.196	15.327	1189.6
260	0.2	0.18627	2782.1	3855.8	51.838	11.074	15.205	1214.9
270	0.2	0.17937	2892.3	4007.3	52.41	10.966	15.096	1239.7
280	0.2	0.17297	3001.5	4157.8	52.957	10.871	15.001	1263.9
290	0.2	0.16701	3109.8	4307.4	53.482	10.79	14.919	1287.5
300	0.2	0.16145	3217.4	4456.2	53.987	10.72	14.848	1310.7
400	0.2	0.12111	4270.3	5921.7	58.205	10.426	14.553	1518.9
500	0.2	0.096905	5311.1	7374.9	61.447	10.402	14.527	1698.4
600	0.2	0.080764	6352.2	8828.5	64.098	10.425	14.55	1859.7
700	0.2	0.069233	7397.1	10.286	66.344	10.479	14.604	2007
800	0.2	0.060583	8449.4	11.751	68.3	10.573	14.697	2142.7
900	0.2	0.053855	9513	13.227	70.038	10.705	14.83	2268.6
1000	0.2	0.048472	10.592	14.718	71.609	10.871	14.995	2386.2
0.5 MPa								
13.963	0.5	77.262	−53.632	−47.161	−3.0705	5.1811	6.9229	1271.9
20	0.5	71.671	−5.7852	1.191	−0.21871	5.6358	9.3847	1139.7
27.112	0.5	60.79	79.588	87.813	3.4355	6.1989	16.637	858.25
27.112	0.5	6.1364	375.87	457.35	17.066	7.0422	19.071	377.59
30	0.5	4.986	404.24	504.52	18.722	6.537	14.608	417.86
40	0.5	3.2933	480.12	631.94	22.406	6.3094	11.791	510.19
50	0.5	2.5276	548.51	746.32	24.961	6.3413	11.206	579.03

(continued)

Table 4.6 (continued)

Temperature (K)	Pressure (MPa)	Density (kg/m³)	Internal energy (kJ/kg)	Enthalpy (kJ/kg)	Entropy (kJ/kg K)	C_v (kJ/kg K)	C_p (kJ/kg K)	Sound speed (m/s)
60	0.5	2.0659	615.81	857.71	26.992	6.5423	11.138	634.57
70	0.5	1.754	685.18	970.24	28.726	6.9697	11.42	679.2
80	0.5	1.5259	759.38	1087.1	30.285	7.6192	11.982	715.45
90	0.5	1.3516	840.57	1210.5	31.738	8.4252	12.732	746.48
100	0.5	1.2137	929.96	1341.9	33.122	9.2907	13.558	775
110	0.5	1.1017	1027.7	1481.6	34.452	10.12	14.36	802.68
120	0.5	1.0088	1133.2	1628.8	35.733	10.842	15.061	830.35
130	0.5	0.9306	1245	1782.3	36.961	11.415	15.618	858.3
140	0.5	0.86372	1361.7	1940.6	38.134	11.827	16.018	886.58
150	0.5	0.80588	1481.7	2102.2	39.249	12.088	16.269	915.12
160	0.5	0.75535	1603.6	2265.6	40.303	12.218	16.391	943.8
170	0.5	0.71082	1726.3	2429.7	41.298	12.244	16.411	972.53
180	0.5	0.67128	1848.7	2593.5	42.235	12.193	16.354	1001.2
190	0.5	0.63592	1970.3	2756.6	43.116	12.089	16.246	1029.6
200	0.5	0.60411	2090.7	2918.3	43.946	11.952	16.105	1057.8
210	0.5	0.57535	2209.6	3078.6	44.728	11.798	15.948	1085.7
220	0.5	0.54921	2326.9	3237.3	45.466	11.639	15.786	1113.1
230	0.5	0.52535	2442.6	3394.3	46.164	11.483	15.627	1140
240	0.5	0.50348	2556.8	3549.9	46.826	11.335	15.478	1166.4
250	0.5	0.48336	2669.5	3703.9	47.455	11.199	15.34	1192.2
260	0.5	0.46479	2781	3856.7	48.055	11.077	15.216	1217.5
270	0.5	0.4475	2891.3	4008.3	48.627	10.969	15.107	1242.3
280	0.5	0.43164	3000.5	4158.9	49.174	10.874	15.011	1266.4
290	0.5	0.41678	3108.9	4308.6	49.7	10.792	14.928	1290.1
300	0.5	0.40291	3216.5	4457.5	50.204	10.722	14.857	1313.2
400	0.5	0.30235	4269.8	5923.6	54.424	10.428	14.557	1521.2
500	0.5	0.24198	5310.8	7377.1	57.668	10.403	14.529	1700.5
600	0.5	0.20171	6352.1	8830.9	60.318	10.426	14.551	1861.5
700	0.5	0.17293	7397.1	10.288	62.565	10.48	14.605	2008.7
800	0.5	0.15135	8449.5	11.753	64.521	10.574	14.698	2144.3
900	0.5	0.13455	9513.1	13.229	66.259	10.706	14.83	2270.1
1000	0.5	0.12111	10.592	14.720	67.83	10.871	14.995	2387.5
1.0 MPa								
14.126	1	77.544	−53.375	−40.48	−3.0547	5.218	6.9169	1281.8
20	1	72.292	−7.7867	6.0461	−0.32325	5.6336	9.1868	1164
30	1	55.303	122.83	140.91	4.992	6.4108	23.144	737.65
31.244	1	49.649	157.25	177.39	6.1813	6.7176	40.772	596.53
31.244	1	14.307	344.36	414.26	13.762	8.0169	52.898	374.7
40	1	7.27	453	590.55	18.848	6.4549	14.122	497.22
50	1	5.2796	530.27	719.68	21.737	6.4095	12.17	574.87
60	1	4.2279	601.87	838.39	23.903	6.5878	11.698	634.16
70	1	3.5511	673.89	955.49	25.708	7.0027	11.794	680.67
80	1	3.0716	749.92	1075.5	27.309	7.6445	12.251	717.92
90	1	2.7113	832.44	1201.3	28.79	8.4455	12.935	749.52

(continued)

Table 4.6 (continued)

Temperature (K)	Pressure (MPa)	Density (kg/m³)	Internal energy (kJ/kg)	Enthalpy (kJ/kg)	Entropy (kJ/kg K)	C_v (kJ/kg K)	C_p (kJ/kg K)	Sound speed (m/s)
100	1	2.4293	922.86	1334.5	30.193	9.3075	13.717	778.39
110	1	2.202	1021.4	1475.6	31.537	10.134	14.488	806.33
120	1	2.0145	1127.6	1623.9	32.828	10.854	15.166	834.2
130	1	1.8571	1240	1778.4	34.064	11.426	15.706	862.31
140	1	1.7229	1357.1	1937.6	35.243	11.837	16.093	890.71
150	1	1.607	1477.5	2099.8	36.362	12.097	16.333	919.34
160	1	1.506	1599.8	2263.8	37.421	12.227	16.447	948.11
170	1	1.417	1722.7	2428.4	38.419	12.252	16.46	976.89
180	1	1.3381	1845.4	2592.7	39.358	12.2	16.397	1005.6
190	1	1.2676	1967.3	2756.2	40.242	12.096	16.284	1034.1
200	1	1.2042	2087.9	2918.3	41.073	11.958	16.139	1062.3
210	1	1.1468	2206.9	3078.9	41.857	11.804	15.978	1090.1
220	1	1.0948	2324.4	3237.9	42.596	11.644	15.813	1117.5
230	1	1.0472	2440.3	3395.2	43.296	11.488	15.652	1144.4
240	1	1.0037	2554.6	3550.9	43.959	11.34	15.5	1170.8
250	1	0.96364	2667.5	3705.2	44.588	11.204	15.361	1196.6
260	1	0.92667	2779.1	3858.2	45.188	11.082	15.235	1221.9
270	1	0.89245	2889.5	4010	45.761	10.973	15.124	1246.6
280	1	0.86067	2998.9	4160.7	46.31	10.879	15.027	1270.7
290	1	0.83109	3107.3	4310.6	46.835	10.797	14.943	1294.3
300	1	0.80348	3215	4459.6	47.341	10.726	14.87	1317.4
400	1	0.60325	4269	5926.7	51.563	10.431	14.564	1524.9
500	1	0.483	5310.4	7380.8	54.808	10.406	14.533	1703.8
600	1	0.40275	6352	8834.9	57.459	10.428	14.554	1864.6
700	1	0.34538	7397.2	10.293	59.706	10.482	14.606	2011.5
800	1	0.30231	8449.7	11.757	61.662	10.575	14.699	2146.9
900	1	0.2688	9513.4	13.234	63.401	10.707	14.831	2272.5
1000	1	0.24198	10.592	14.725	64.971	10.873	14.996	2389.7
2.0 MPa								
14.447	2	78.094	−52.783	−27.172	−3.0227	5.2606	6.8932	1303.7
20	2	73.43	−11.278	15.959	−0.51378	5.6274	8.8607	1208.3
30	2	59.822	102.02	135.45	4.2332	6.3219	16.572	887.98
40	2	19.189	377.15	481.38	13.981	6.9549	26.646	485.62
50	2	11.546	490.53	663.75	18.094	6.5403	14.668	574.31
60	2	8.822	573.12	799.83	20.579	6.6684	12.929	638.08
70	2	7.2589	651.1	926.63	22.535	7.0621	12.564	686.73
80	2	6.2105	731	1053	24.222	7.691	12.789	725.03
90	2	5.447	816.29	1183.5	25.758	8.4833	13.335	757.17
100	2	4.8611	908.79	1320.2	27.198	9.3392	14.028	786.37
110	2	4.395	1009	1464.1	28.569	10.162	14.737	814.54
120	2	4.014	1116.5	1614.7	29.879	10.878	15.37	842.6
130	2	3.6961	1230	1771.1	31.13	11.447	15.877	870.87
140	2	3.4264	1348.1	1931.8	32.321	11.857	16.238	899.41
150	2	3.1944	1469.2	2095.3	33.449	12.115	16.458	928.15

(continued)

Table 4.6 (continued)

Temperature (K)	Pressure (MPa)	Density (kg/m^3)	Internal energy (kJ/kg)	Enthalpy (kJ/kg)	Entropy (kJ/kg K)	C_v (kJ/kg K)	C_p (kJ/kg K)	Sound speed (m/s)
160	2	2.9925	1592.2	2260.5	34.515	12.243	16.555	957
170	2	2.8152	1715.7	2426.1	35.519	12.267	16.554	985.84
180	2	2.6531	1838.9	2591.3	36.464	12.215	16.481	1014.6
190	2	2.5179	1961.3	2755.6	37.352	12.109	16.358	1043.1
200	2	2.392	2082.3	2918.4	38.187	11.971	16.205	1071.3
210	2	2.2732	2201.8	3079.6	38.974	11.816	16.038	1099.1
220	2	2.1749	2319.6	3239.2	39.716	11.656	15.867	1126.5
230	2	2.0807	2435.8	3397	40.417	11.499	15.701	1153.3
240	2	1.9943	2550.4	3553.2	41.082	11.351	15.545	1179.7
250	2	1.915	2663.5	3707.9	41.714	11.214	15.402	1205.4
260	2	1.8417	2775.3	3861.3	42.315	11.091	15.273	1230.6
270	2	1.7739	2886	4013.4	42.89	10.983	15.158	1255.2
280	2	1.7109	2995.6	4164.5	43.439	10.887	15.058	1279.3
290	2	1.6523	3104.2	4314.6	43.966	10.805	14.972	1302.8
300	2	1.5976	3212.1	4464	44.472	10.735	14.897	1325.8
400	2	1.2008	4267.5	5933	48.7	10.437	14.578	1532.5
500	2	0.96222	5309.6	7388.2	51.948	10.411	14.541	1710.6
600	2	0.80284	6351.7	8842.9	54.6	10.432	14.558	1870.7
700	2	0.68879	7397.3	10.301	56.847	10.486	14.609	2017.1
800	2	0.60313	8450.1	11.766	58.804	10.578	14.701	2152
900	2	0.53643	9514	13.242	60.542	10.71	14.832	2277.3
1000	2	0.48302	10.593	14.734	62.113	10.875	14.996	2394.2
5.0 MPa								
15.371	5	79.627	−50.555	12.237	−2.9305	5.2627	6.7891	1377.3
20	5	76.273	−18.92	46.634	−0.98265	5.5963	8.2007	1319.9
30	5	66.452	73.637	148.88	3.1054	6.3503	12.46	1109
40	5	51.392	206.48	303.77	7.5163	6.5737	18.947	839.51
50	5	33.684	363.65	512.09	12.158	6.7238	20.577	691.56
60	5	23.784	485.69	695.91	15.521	6.8387	16.49	700.83
70	5	18.768	583.9	850.31	17.904	7.1994	14.707	734.31
80	5	15.707	676.03	994.35	19.828	7.8054	14.245	765.69
90	5	13.603	769.67	1137.2	21.511	8.5809	14.405	793.78
100	5	12.046	868.33	1283.4	23.051	9.424	14.856	820.39
110	5	10.838	973.32	1434.7	24.492	10.237	15.4	846.87
120	5	9.8675	1084.6	1591.3	25.854	10.945	15.915	873.8
130	5	9.0678	1201.2	1752.6	27.146	11.508	16.333	901.32
140	5	8.3954	1322	1917.5	28.368	11.912	16.626	929.32
150	5	7.821	1445.4	2084.7	29.521	12.166	16.793	957.68
160	5	7.3238	1570.3	2253	30.607	12.29	16.847	986.21
170	5	6.8887	1695.5	2421.4	31.628	12.311	16.811	1014.8
180	5	6.5043	1820.3	2589	32.586	12.256	16.708	1043.3
190	5	6.1621	1944	2755.4	33.486	12.148	16.561	1071.5
200	5	5.8552	2066.2	2920.2	34.331	12.008	16.387	1099.5
210	5	5.5783	2186.8	3083.1	35.126	11.851	16.202	1127.1

(continued)

Table 4.6 (continued)

Temperature (K)	Pressure (MPa)	Density (kg/m^3)	Internal energy (kJ/kg)	Enthalpy (kJ/kg)	Entropy (kJ/kg K)	C_v (kJ/kg K)	C_p (kJ/kg K)	Sound speed (m/s)
220	5	5.327	2305.6	3244.2	35.875	11.689	16.016	1154.2
230	5	5.098	2422.7	3403.4	36.583	11.531	15.836	1180.8
240	5	4.8883	2538.1	3561	37.254	11.381	15.668	1206.9
250	5	4.6956	2652	3716.9	37.89	11.244	15.515	1232.3
260	5	4.5177	2764.6	3871.3	38.496	11.12	15.377	1257.2
270	5	4.3531	2875.8	4024.4	39.074	11.01	15.254	1281.6
280	5	4.2002	2986	4176.4	39.627	10.913	15.147	1305.3
290	5	4.0579	3095.3	4327.4	40.156	10.83	15.054	1328.6
300	5	3.925	3203.7	4477.6	40.665	10.759	14.974	1351.3
400	5	2.9599	4262.9	5952.1	44.91	10.456	14.617	1555.1
500	5	2.3776	5307.4	7410.3	48.164	10.425	14.564	1730.8
600	5	1.9874	6351	8866.8	50.819	10.444	14.572	1889
700	5	1.7075	7397.7	10.326	53.069	10.496	14.618	2033.8
800	5	1.4967	8451.3	11.792	55.026	10.587	14.706	2167.5
900	5	1.3324	9515.9	13.269	56.765	10.718	14.835	2291.6
1000	5	1.2005	10.595	14.760	58.336	10.882	14.998	2407.6
10.0 MPa								
16.804	10	81.857	−45.94	76.224	−2.8011	5.2146	6.6757	1498.2
20	10	79.907	−26.226	98.919	−1.5669	5.5295	7.5665	1466.8
30	10	72.488	52.061	190.02	2.0849	6.3939	10.615	1321.3
40	10	63.191	152.25	310.51	5.5316	6.6645	13.447	1153.7
50	10	52.666	266.92	456.8	8.787	6.7648	15.598	1007.8
60	10	42.864	383.73	617.03	11.707	6.9603	16.193	923.25
70	10	35.294	494.07	777.4	14.18	7.3491	15.82	890.74
80	10	29.873	598.86	933.62	16.266	7.9516	15.474	884.94
90	10	25.951	702.76	1088.1	18.086	8.7155	15.481	891.32
100	10	23.007	809.5	1244.2	19.73	9.5465	15.76	904.33
110	10	20.714	920.95	1403.7	21.25	10.348	16.161	921.78
120	10	18.872	1037.5	1567.4	22.674	11.047	16.562	942.45
130	10	17.356	1158.5	1734.7	24.013	11.602	16.89	965.45
140	10	16.083	1283	1904.8	25.274	11.999	17.11	990.11
150	10	14.997	1409.7	2076.5	26.458	12.247	17.217	1015.9
160	10	14.057	1537.4	2248.8	27.57	12.366	17.223	1042.4
170	10	13.235	1665.1	2420.7	28.612	12.382	17.146	1069.3
180	10	12.508	1792.1	2591.5	29.589	12.323	17.009	1096.3
190	10	11.862	1917.7	2760.8	30.504	12.211	16.832	1123.4
200	10	11.282	2041.7	2928.1	31.362	12.068	16.633	1150.3
210	10	10.759	2163.9	3093.4	32.169	11.908	16.426	1176.9
220	10	10.284	2284.2	3256.6	32.928	11.744	16.22	1203.1
230	10	9.8504	2402.6	3417.8	33.645	11.583	16.024	1228.9
240	10	9.4534	2519.3	3577.1	34.323	11.431	15.841	1254.1
250	10	9.0883	2634.4	3734.7	34.966	11.291	15.674	1278.8
260	10	8.7512	2748	3890.7	35.578	11.166	15.524	1303
270	10	8.4389	2860.2	4045.2	36.161	11.054	15.39	1326.6

(continued)

Table 4.6 (continued)

Temperature (K)	Pressure (MPa)	Density (kg/m^3)	Internal energy (kJ/kg)	Enthalpy (kJ/kg)	Entropy (kJ/kg K)	C_v (kJ/kg K)	C_p (kJ/kg K)	Sound speed (m/s)
280	10	8.1487	2971.3	4198.5	36.718	10.956	15.273	1349.7
290	10	7.8784	3081.4	4350.7	37.253	10.871	15.172	1372.3
300	10	7.6258	3190.7	4502	37.765	10.799	15.083	1394.4
400	10	5.784	4255.8	5984.7	42.034	10.485	14.677	1592.7
500	10	4.6652	5303.9	7447.4	45.298	10.449	14.599	1764.4
600	10	3.9112	6350	8906.8	47.959	10.464	14.594	1919.3
700	10	3.3679	7398.5	10.368	50.211	10.512	14.632	2061.5
800	10	2.9574	8453.4	11.835	52.169	10.601	14.715	2193
900	10	2.6363	9519	13.312	53.909	10.73	14.841	2315.3
1000	10	2.3781	10.599	14.804	55.481	10.893	15.002	2429.7
50.0 MPa								
25.707	50	93.4	8.6495	543.98	−2.1536	5.8331	7.0604	2057.5
30	50	91.852	31.818	576.17	−0.99724	6.273	7.9171	2040
40	50	87.893	94.717	663.59	1.5069	6.9196	9.4683	1996.3
50	50	83.72	166.78	764.01	3.7432	7.2738	10.563	1946.7
60	50	79.536	245.55	874.2	5.7497	7.5971	11.461	1890.7
70	50	75.456	330.55	993.19	7.5821	8.048	12.343	1829.2
80	50	71.551	422.43	1121.2	9.2906	8.6731	13.277	1766.2
90	50	67.862	522.06	1258.9	10.91	9.4332	14.247	1707.7
100	50	64.412	629.84	1406.1	12.461	10.245	15.189	1658.2
110	50	61.21	745.46	1562.3	13.949	11.019	16.035	1619.4
120	50	58.252	867.95	1726.3	15.375	11.687	16.732	1591.3
130	50	55.526	995.89	1896.4	16.737	12.211	17.254	1572.7
140	50	53.018	1127.7	2070.8	18.029	12.577	17.6	1561.9
150	50	50.71	1261.8	2247.8	19.25	12.796	17.784	1557.5
160	50	48.585	1396.9	2426	20.4	12.887	17.834	1558.1
170	50	46.625	1531.8	2604.2	21.48	12.878	17.779	1562.8
180	50	44.815	1665.7	2781.4	22.493	12.795	17.648	1570.5
190	50	43.14	1797.9	2957	23.443	12.662	17.467	1580.6
200	50	41.587	1928.3	3130.6	24.333	12.498	17.256	1592.5
210	50	40.143	2056.5	3302	25.17	12.319	17.032	1605.6
220	50	38.798	2182.5	3471.2	25.957	12.138	16.808	1619.7
230	50	37.543	2306.4	3638.2	26.699	11.961	16.59	1634.4
240	50	36.368	2428.2	3803.1	27.401	11.794	16.385	1649.5
250	50	35.267	2548.2	3966	28.066	11.641	16.196	1664.9
260	50	34.233	2666.5	4127.1	28.698	11.502	16.024	1680.4
270	50	33.26	2783.2	4286.5	29.3	11.379	15.869	1695.9
280	50	32.342	2898.5	4444.5	29.874	11.27	15.731	1711.4
290	50	31.476	3012.7	4601.2	30.424	11.175	15.609	1726.8
300	50	30.656	3125.7	4756.7	30.951	11.093	15.501	1742.1
400	50	24.37	4221.3	6273	35.317	10.707	14.943	1888
500	50	20.264	5290.1	7757.5	38.63	10.627	14.775	2024.3
600	50	17.36	6351.1	9231.3	41.318	10.612	14.712	2153.2
700	50	15.192	7410.7	10.702	43.585	10.638	14.712	2275

(continued)

Table 4.6 (continued)

Temperature (K)	Pressure (MPa)	Density (kg/m^3)	Internal energy (kJ/kg)	Enthalpy (kJ/kg)	Entropy (kJ/kg K)	C_v (kJ/kg K)	C_p (kJ/kg K)	Sound speed (m/s)
800	50	13.509	8474.3	12.176	45.552	10.711	14.769	2389.8
900	50	12.164	9546.9	13.657	47.298	10.827	14.876	2498.2
1000	50	11.063	10.633	15.152	48.872	10.979	15.024	2600.8
100.0 MPa								
34.181	100	102.49	92.366	1068	−1.6612	6.2333	7.4643	2480.7
40	100	100.85	122.72	1114.3	−0.41346	6.7537	8.3979	2471.3
50	100	97.841	182.59	1204.7	1.5979	7.3572	9.6104	2447.4
60	100	94.76	250.4	1305.7	3.4374	7.843	10.578	2414.2
70	100	91.708	325.7	1416.1	5.1378	8.3992	11.511	2371.7
80	100	88.735	409.16	1536.1	6.7385	9.0915	12.496	2322.6
90	100	85.869	501.65	1666.2	8.2696	9.8927	13.525	2271.9
100	100	83.123	603.52	1806.6	9.7475	10.727	14.537	2224.3
110	100	80.5	714.41	1956.6	11.177	11.512	15.458	2182.7
120	100	78.003	833.23	2115.2	12.557	12.183	16.233	2148.1
130	100	75.628	958.46	2280.7	13.881	12.702	16.833	2120.5
140	100	73.373	1088.4	2451.3	15.145	13.061	17.251	2099.4
150	100	71.232	1221.3	2625.2	16.344	13.27	17.504	2083.8
160	100	69.2	1355.8	2800.9	17.478	13.351	17.615	2073
170	100	67.271	1490.6	2977.1	18.547	13.329	17.615	2066.2
180	100	65.44	1624.8	3152.9	19.552	13.234	17.532	2062.7
190	100	63.7	1757.7	3327.6	20.496	13.088	17.393	2061.8
200	100	62.047	1889	3500.7	21.384	12.911	17.219	2063.2
210	100	60.475	2018.3	3671.9	22.219	12.721	17.026	2066.2
220	100	58.979	2145.6	3841.2	23.007	12.527	16.828	2070.6
230	100	57.554	2271	4008.5	23.75	12.339	16.633	2076.1
240	100	56.195	2394.4	4173.9	24.454	12.161	16.447	2082.4
250	100	54.9	2516	4337.5	25.122	11.997	16.274	2089.3
260	100	53.663	2635.9	4499.4	25.757	11.848	16.115	2096.8
270	100	52.48	2754.3	4659.8	26.363	11.715	15.97	2104.6
280	100	51.35	2871.4	4818.8	26.941	11.596	15.841	2112.7
290	100	50.267	2987.3	4976.7	27.495	11.492	15.725	2121
300	100	49.23	3102.1	5133.4	28.026	11.401	15.623	2129.6
400	100	40.846	4214.3	6662.6	32.429	10.949	15.069	2221.5
500	100	34.948	5297.1	8158.5	35.768	10.823	14.876	2320.4
600	100	30.565	6369.3	9641	38.471	10.777	14.788	2421.4
700	100	27.173	7438.1	11.118	40.749	10.781	14.767	2521.3
800	100	24.466	8509.2	12.596	42.722	10.835	14.808	2618.2
900	100	22.254	9588	14.082	44.472	10.938	14.903	2711.4
1000	100	20.411	10.679	15.578	46.048	11.078	15.04	2801

4.4 Formulation for Normal Hydrogen

The functional form for the parahydrogen EOS was refit to the normal hydrogen surface of state with some transformation of parahydrogen property measurements that use the Quantum Law of Corresponding States. As such, differences between the parahydrogen and normal hydrogen surfaces of state are small. The uncertainty statement in REFPROP for this formulation follows.

The uncertainty in density is 0.1% at temperatures from the triple point to 250 K and at pressures up to 40 MPa, except in the critical region, where an uncertainty of 0.2% in pressure is generally attained. In the region between 250 and 450 K and at pressures to 300 MPa, the uncertainty in density is 0.04%. At temperatures between 450 and 1000 K, the uncertainty in density increases to 1%. At pressures between 300 and 2000 MPa, the uncertainty in density is 8%. Speed of sound data are represented within 0.5% below 100 MPa. The estimated uncertainty for heat capacities is 1.0%. The estimated uncertainties of vapor pressures and saturated liquid densities calculated with the Maxwell criterion are 0.2% for each property (Tables 4.7, 4.8, 4.9 and 4.10).

Table 4.7 Fixed point properties and correlation limits for normal hydrogen

	Temperature (K)	Pressure (kPa)	Density (mol/dm^3)
Critical point	33.145	1296.4	15.508
Triple point	13.957	7.36	
Upper limit	1000	2,000,000	102.0
Lower limit	13.957		

Table 4.8 Fluid constants and normal boiling point for normal hydrogen

Molar mass	2.01588 g/mol
Ideal-gas constant	8.314472 J/(mol K)
Pitzer's acentric factor	−0.219
Normal boiling point	20.369 K

Table 4.9 Fundamental equation of state coefficients and exponents for normal hydrogen

i	N_i	t_i	d_i	p_i	φ_i	β_i	γ_i	D_i
1	−6.93643	0.6844	1					
2	0.01	1	4					
3	2.1101	0.989	1					
4	4.52059	0.489	1					
5	0.732564	0.803	2					

(continued)

Table 4.9 (continued)

i	N_i	t_i	d_i	p_i	φ_i	β_i	γ_i	D_i
6	−1.34086	1.1444	2					
7	0.130985	1.409	3					
8	−0.777414	1.754	1	1				
9	0.351944	1.311	3	1				
10	−0.0211716	4.187	2		1.685	0.171	0.7164	1.506
11	0.0226312	5.646	1		0.489	0.2245	1.3444	0.156
12	0.032187	0.791	3		0.103	0.1304	1.4517	1.736
13	−0.0231752	7.249	1		2.506	0.2785	0.7204	0.67
14	0.0557346	2.986	1		1.607	0.3967	1.5445	1.662

Table 4.10 Ancillary function coefficients and exponents for normal hydrogen

	α^0	
i	a_i	k_i
0	2.5	
1	−1.4579856475	
2	1.888076782	
3	1.616	531.0
4	−0.4117	751.0
5	−0.792	1989.0
6	0.758	2484.0
7	1.217	6859.0

$m = 0$, $n = 5$

	Vapor pressure		Saturated liquid density		Saturated vapor density	
i	N_i	k_i	N_i	k_i	N_i	k_i
1	−4.8979	1	15.456	0.62	−2.9962	0.466
2	0.988558	1.5	−41.72	0.83	−16.724	2
3	0.349689	2	50.276	1.05	15.819	2.4
4	0.499356	2.85	−27.947	1.3	−16.852	4
5			5.6718	1.6	34.586	7
6					−53.754	8

4.5 Property Plots for Normal Hydrogen

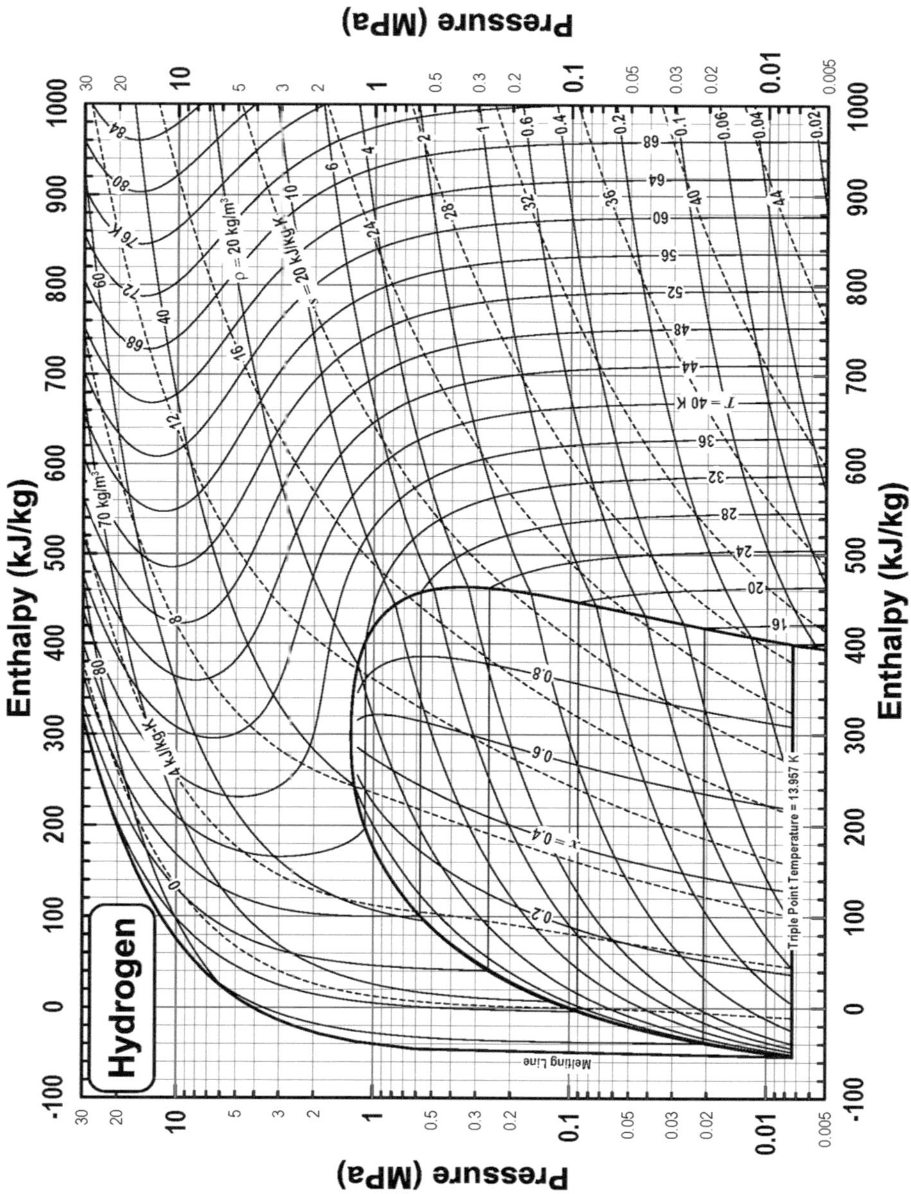
Hydrogen
Enthalpy (kJ/kg)
Pressure (MPa)
Melting Line
Triple Point Temperature = 13.957 K
T = 40 K
ρ = 20 kg/m³
s = 20 kJ/kg-K
x = 0.4

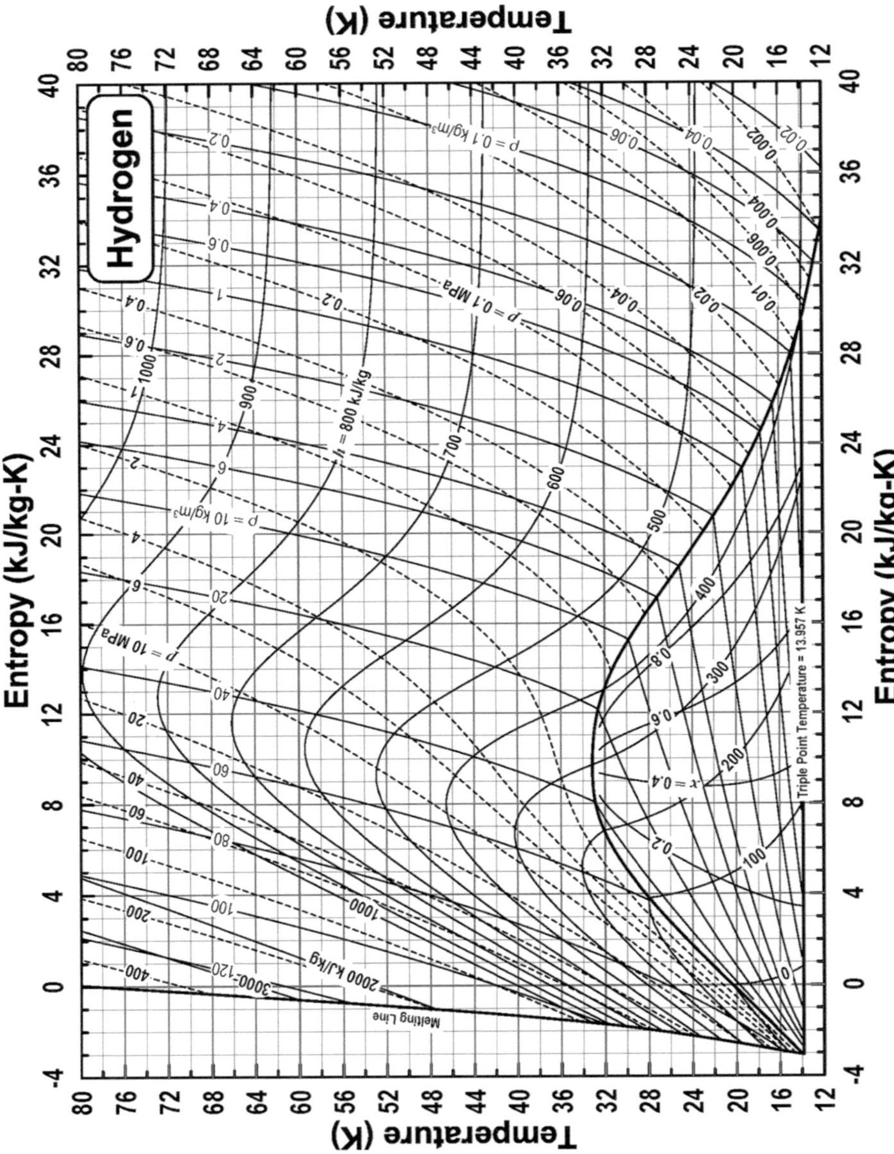
Hydrogen
Entropy (kJ/kg-K)
Temperature (K)
Melting Line
Triple Point Temperature = 13.957 K
p = 10 MPa
p = 0.1 MPa
ρ = 10 kg/m³
ρ = 0.1 kg/m³
h = 800 kJ/kg
2000 kJ/kg
x = 0.4

4.6 Thermodynamic Property Tables for Normal Hydrogen

See Tables 4.11 and 4.12.

Table 4.11 Thermodynamic properties of saturated normal hydrogen

Temperature (K)	Pressure (MPa)	Liquid density (kg/m^3)	Vapor density (kg/m^3)	Liquid enthalpy (kJ/kg)	Vapor enthalpy (kJ/kg)	Liquid entropy (kJ/kg K)	Vapor entropy (kJ/kg K)	Liquid C_v (kJ/kg K)	Vapor C_v (kJ/kg K)	Liquid sound speed (m/s)	Vapor sound speed (m/s)
13.957	0.007358	77.004	0.12985	−53.926	399.83	−3.0723	29.438	5.1616	6.2433	7.0212	10.564
14	0.007541	76.969	0.13272	−53.622	400.22	−3.0507	29.367	5.1625	6.2449	7.0312	10.57
15	0.012898	76.136	0.21346	−46.388	409.28	−2.5566	27.821	5.2313	6.2814	7.3386	10.703
16	0.020755	75.264	0.32506	−38.777	417.91	−2.0722	26.471	5.3287	6.314	7.7212	10.86
17	0.031759	74.345	0.47356	−30.733	426.05	−1.5936	25.276	5.4192	6.3427	8.1368	11.047
18	0.046602	73.375	0.66527	−22.224	433.6	−1.1188	24.204	5.4986	6.3701	8.5793	11.274
19	0.066006	72.35	0.90693	−13.215	440.49	−0.64625	23.233	5.57	6.3996	9.0539	11.551
20	0.090717	71.265	1.2059	−3.6672	446.64	−0.17429	22.341	5.6369	6.4343	9.5697	11.892
20.369	0.10133	70.848	1.3322	0	448.71	0	22.029	5.6609	6.449	9.7725	12.037
21	0.1215	70.115	1.5701	6.466	451.98	0.29876	21.514	5.7015	6.4768	10.138	12.312
22	0.15913	68.893	2.009	17.241	456.43	0.77472	20.738	5.7649	6.5292	10.771	12.83
23	0.20438	67.592	2.5334	28.724	459.88	1.2556	20.002	5.8276	6.5927	11.49	13.472
24	0.25807	66.199	3.1562	40.997	462.24	1.7437	19.296	5.8898	6.6685	12.319	14.273
25	0.321	64.701	3.8938	54.161	463.37	2.2417	18.61	5.9521	6.7578	13.298	15.289
26	0.39399	63.079	4.7674	68.346	463.1	2.7531	17.936	6.0155	6.862	14.487	16.603
27	0.47789	61.305	5.8055	83.727	461.18	3.2825	17.262	6.082	6.9832	15.987	18.357
28	0.57359	59.339	7.0489	100.55	457.3	3.8365	16.577	6.1552	7.1248	17.977	20.804
29	0.68205	57.119	8.5601	119.2	450.92	4.4253	15.864	6.2415	7.2922	20.807	24.445
30	0.80432	54.538	10.445	140.3	441.19	5.0661	15.096	6.3535	7.4945	25.284	30.425
31	0.94165	51.381	12.909	165.08	426.41	5.7933	14.223	6.5181	7.7475	33.759	42.069
32	1.0957	47.085	16.495	196.73	402.3	6.6983	13.122	6.811	8.0826	57.287	74.636
33	1.2693	38.079	24.637	255.69	343.4	8.3842	11.042	7.6982	8.5381	484.58	604.72
33.145	1.2965	31.262	31.262	298.16	298.16	9.6442	9.6442				

Table 4.12 Thermodynamic properties of normal hydrogen

Temperature (K)	Pressure (MPa)	Density (kg/m³)	Internal energy (kJ/kg)	Enthalpy (kJ/kg)	Entropy (kJ/kg K)	C_v (kJ/kg K)	C_p (kJ/kg K)	Sound speed (m/s)
0.1 MPa								
13.957	0.007358	77.004	−54.022	−53.926	−3.0723	5.1616	7.0212	1269.2
20	0.1	71.277	−4.9821	−3.5791	−0.1764	5.637	9.5654	1129.6
20.324	0.1	70.899	−1.8572	−0.44679	−0.02104	5.6581	9.7476	1120.1
20.324	0.1	1.3164	372.51	448.47	22.066	6.4471	12.018	356.17
30	0.1	0.83596	437.25	556.87	26.443	6.2332	10.818	447.82
40	0.1	0.6155	501.13	663.6	29.515	6.2167	10.57	521.54
50	0.1	0.48878	564.19	768.78	31.862	6.222	10.482	584.75
60	0.1	0.40587	627.17	873.55	33.772	6.2716	10.485	640.4
70	0.1	0.34722	690.77	978.77	35.394	6.3849	10.572	689.7
80	0.1	0.30348	755.74	1085.3	36.816	6.5655	10.736	733.56
90	0.1	0.26957	822.74	1193.7	38.093	6.802	10.962	772.95
100	0.1	0.2425	892.26	1304.6	39.261	7.0757	11.228	808.85
110	0.1	0.22039	964.6	1418.3	40.345	7.3677	11.515	842.14
120	0.1	0.20199	1039.9	1534.9	41.359	7.6622	11.805	873.5
130	0.1	0.18642	1118	1654.4	42.315	7.9482	12.088	903.41
140	0.1	0.17309	1198.9	1776.6	43.221	8.2185	12.356	932.21
150	0.1	0.16154	1282.4	1901.5	44.082	8.4692	12.605	960.13
160	0.1	0.15144	1368.3	2028.7	44.903	8.6984	12.832	987.3
170	0.1	0.14253	1456.4	2158	45.687	8.9061	13.039	1013.8
180	0.1	0.13461	1546.5	2289.4	46.438	9.0928	13.225	1039.8
190	0.1	0.12752	1638.3	2422.5	47.157	9.2597	13.391	1065.3
200	0.1	0.12115	1731.7	2557.1	47.848	9.4082	13.538	1090.2
210	0.1	0.11538	1826.5	2693.2	48.512	9.5398	13.669	1114.8
220	0.1	0.11013	1922.5	2830.5	49.151	9.656	13.785	1138.9
230	0.1	0.10535	2019.6	2968.8	49.766	9.7583	13.887	1162.6
240	0.1	0.10096	2117.6	3108.2	50.359	9.848	13.976	1186
250	0.1	0.09692	2216.5	3248.3	50.931	9.9263	14.054	1209
260	0.1	0.093193	2316.2	3389.2	51.483	9.9945	14.122	1231.7
270	0.1	0.089742	2416.4	3530.7	52.017	10.054	14.181	1254.1
280	0.1	0.086538	2517.2	3672.8	52.534	10.105	14.232	1276.1
290	0.1	0.083555	2618.5	3815.3	53.034	10.149	14.275	1297.9
300	0.1	0.080771	2720.2	3958.3	53.519	10.186	14.313	1319.3
400	0.1	0.060585	3749.6	5400.2	57.666	10.354	14.479	1519.6
500	0.1	0.048472	4787	6850.1	60.901	10.388	14.513	1698
600	0.1	0.040396	5827.5	8303	63.55	10.425	14.549	1859.1
700	0.1	0.034626	6872.9	9760.9	65.797	10.49	14.614	2006.2
800	0.1	0.030299	7926.4	11.227	67.754	10.585	14.71	2141.8
900	0.1	0.026933	8991	12.704	69.494	10.712	14.836	2267.9
1000	0.1	0.024241	10.070	14.195	71.065	10.868	14.992	2385.8
0.2 MPa								
14.022	0.2	77.113	−53.935	−51.342	−3.0663	5.1676	7.0132	1273.9
20	0.2	71.409	−5.4283	−2.6276	−0.1989	5.638	9.5201	1134.7
22.91	0.2	67.712	24.711	27.665	1.2123	5.822	11.421	1040.5
22.91	0.2	2.4826	379.06	459.62	20.066	6.5865	13.408	368.53

(continued)

Table 4.12 (continued)

Temperature (K)	Pressure (MPa)	Density (kg/m^3)	Internal energy (kJ/kg)	Enthalpy (kJ/kg)	Entropy (kJ/kg K)	C_v (kJ/kg K)	C_p (kJ/kg K)	Sound speed (m/s)
30	0.2	1.735	430.33	545.6	23.347	6.2912	11.444	441.08
40	0.2	1.2507	496.56	656.47	26.541	6.245	10.848	518.72
50	0.2	0.98549	560.8	763.75	28.935	6.2389	10.641	583.6
60	0.2	0.81535	624.5	869.79	30.869	6.2825	10.588	640.12
70	0.2	0.69617	688.57	975.86	32.504	6.3924	10.643	689.91
80	0.2	0.60777	753.88	1083	33.934	6.5711	10.789	734.06
90	0.2	0.53949	821.14	1191.9	35.216	6.8063	11.002	773.62
100	0.2	0.48511	890.86	1303.1	36.388	7.0792	11.26	809.63
110	0.2	0.44075	963.36	1417.1	37.475	7.3706	11.54	842.99
120	0.2	0.40386	1038.7	1534	38.491	7.6647	11.826	874.39
130	0.2	0.37269	1117	1653.6	39.449	7.9504	12.106	904.33
140	0.2	0.34601	1198	1776	40.356	8.2205	12.371	933.15
150	0.2	0.3229	1281.6	1901	41.218	8.471	12.618	961.07
160	0.2	0.30269	1367.6	2028.3	42.039	8.7001	12.844	988.25
170	0.2	0.28487	1455.7	2157.8	42.824	8.9076	13.049	1014.8
180	0.2	0.26904	1545.8	2289.2	43.575	9.0942	13.233	1040.7
190	0.2	0.25487	1637.7	2422.4	44.295	9.2611	13.398	1066.2
200	0.2	0.24213	1731.1	2557.1	44.987	9.4095	13.545	1091.2
210	0.2	0.2306	1825.9	2693.2	45.651	9.541	13.675	1115.7
220	0.2	0.22012	1922	2830.6	46.289	9.6572	13.79	1139.8
230	0.2	0.21055	2019.1	2969	46.905	9.7594	13.892	1163.6
240	0.2	0.20178	2117.2	3108.4	47.498	9.849	13.981	1186.9
250	0.2	0.19372	2216.1	3248.6	48.07	9.9273	14.058	1209.9
260	0.2	0.18627	2315.8	3389.5	48.623	9.9954	14.126	1232.6
270	0.2	0.17937	2416.1	3531.1	49.157	10.055	14.184	1254.9
280	0.2	0.17297	2516.9	3673.2	49.674	10.106	14.235	1277
290	0.2	0.16701	2618.2	3815.7	50.174	10.15	14.278	1298.7
300	0.2	0.16145	2719.9	3958.7	50.659	10.187	14.316	1320.2
400	0.2	0.12111	3749.5	5400.8	54.807	10.355	14.481	1520.4
500	0.2	0.096905	4786.9	6850.8	58.042	10.389	14.514	1698.7
600	0.2	0.080765	5827.5	8303.8	60.691	10.425	14.55	1859.7
700	0.2	0.069233	6872.9	9761.7	62.938	10.49	14.614	2006.7
800	0.2	0.060584	7926.5	11.228	64.896	10.586	14.71	2142.4
900	0.2	0.053855	8991.1	12.705	66.635	10.712	14.837	2268.4
1000	0.2	0.048472	10.070	14.196	68.206	10.868	14.992	2386.2
0.5 MPa								
14.121	0.5	77.282	−53.797	−47.327	−3.0572	5.1765	7.0016	1281.3
20	0.5	71.795	−6.7143	0.25005	−0.26451	5.6407	9.3924	1149.7
27.243	0.5	60.847	79.45	87.667	3.4144	6.099	16.416	866.9
27.243	0.5	6.0866	378.29	460.44	17.098	7.0155	18.874	379.02
30	0.5	4.9932	405.49	505.63	18.681	6.5937	14.769	417.17
40	0.5	3.2912	482.08	634	22.393	6.3327	11.831	510.2
50	0.5	2.5259	550.39	748.34	24.947	6.2885	11.153	580.37
60	0.5	2.0658	616.38	858.42	26.954	6.3142	10.906	639.47
70	0.5	1.7533	681.95	967.13	28.63	6.4145	10.861	690.69

(continued)

Table 4.12 (continued)

Temperature (K)	Pressure (MPa)	Density (kg/m^3)	Internal energy (kJ/kg)	Enthalpy (kJ/kg)	Entropy (kJ/kg K)	C_v (kJ/kg K)	C_p (kJ/kg K)	Sound speed (m/s)
80	0.5	1.5254	748.31	1076.1	30.085	6.5874	10.947	735.67
90	0.5	1.3512	816.34	1186.4	31.384	6.819	11.123	775.73
100	0.5	1.2134	886.66	1298.7	32.567	7.0895	11.355	812.04
110	0.5	1.1015	959.64	1413.6	33.662	7.3793	11.617	845.59
120	0.5	1.0087	1035.4	1531.1	34.684	7.6722	11.889	877.12
130	0.5	0.93051	1114	1651.3	35.647	7.9571	12.159	907.13
140	0.5	0.86365	1195.3	1774.2	36.557	8.2264	12.416	936
150	0.5	0.80583	1279.1	1899.6	37.422	8.4764	12.656	963.94
160	0.5	0.75531	1365.3	2027.3	38.246	8.7051	12.877	991.12
170	0.5	0.71079	1453.6	2157.1	39.033	8.9122	13.078	1017.7
180	0.5	0.67125	1543.9	2288.8	39.785	9.0985	13.259	1043.6
190	0.5	0.6359	1635.9	2422.2	40.507	9.2651	13.421	1069
200	0.5	0.6041	1729.4	2557.1	41.199	9.4133	13.566	1094
210	0.5	0.57534	1824.4	2693.4	41.864	9.5446	13.694	1118.5
220	0.5	0.5492	1920.5	2831	42.504	9.6606	13.807	1142.6
230	0.5	0.52534	2017.8	2969.5	43.12	9.7626	13.907	1166.3
240	0.5	0.50347	2115.9	3109	43.713	9.8521	13.994	1189.6
250	0.5	0.48336	2214.9	3249.4	44.286	9.9303	14.071	1212.6
260	0.5	0.46479	2314.7	3390.4	44.839	9.9983	14.137	1235.3
270	0.5	0.4476	2415	3532.1	45.374	10.057	14.195	1257.6
280	0.5	0.43164	2515.9	3674.3	45.891	10.108	14.244	1279.6
290	0.5	0.41678	2617.3	3817	46.392	10.152	14.287	1301.3
300	0.5	0.40291	2719	3960	46.877	10.19	14.324	1322.7
400	0.5	0.30235	3749	5402.7	51.026	10.357	14.485	1522.6
500	0.5	0.24198	4786.7	6853	54.262	10.39	14.516	1700.7
600	0.5	0.20171	5827.4	8306.2	56.912	10.426	14.551	1861.5
700	0.5	0.17294	6873	9764.2	59.159	10.491	14.615	2008.4
800	0.5	0.15135	7926.6	11.230	61.117	10.587	14.711	2143.9
900	0.5	0.13455	8991.3	12.707	62.856	10.713	14.837	2269.8
1000	0.5	0.12111	10.070	14.199	64.427	10.868	14.992	2387.6
1.0 MPa								
14.285	1	77.56	−53.553	−40.659	−3.0424	5.1903	6.9845	1293.8
20	1	72.405	−8.6978	5.1133	−0.36807	5.6445	9.203	1173.3
30	1	55.881	119.28	137.17	4.8439	6.3044	22.035	758.02
31.393	1	49.888	156.31	176.36	6.1179	6.6099	39.721	609.07
31.393	1	14.128	347.71	418.49	13.831	7.8666	50.34	377.1
40	1	7.2758	454.58	592.02	18.819	6.4993	14.255	496.6
50	1	5.2763	532.11	721.64	21.72	6.3679	12.145	576.1
60	1	4.2248	602.51	839.21	23.866	6.3645	11.476	639.26
70	1	3.5489	670.77	952.55	25.613	6.4496	11.238	692.63
80	1	3.07	738.95	1064.7	27.111	6.6136	11.216	738.82
90	1	2.7102	808.32	1177.3	28.437	6.8397	11.325	779.59
100	1	2.4285	879.67	1291.4	29.639	7.1064	11.513	816.34
110	1	2.2014	953.45	1407.7	30.747	7.3935	11.744	850.15
120	1	2.0141	1029.9	1526.4	31.78	7.6846	11.994	881.83

(continued)

Table 4.12 (continued)

Temperature (K)	Pressure (MPa)	Density (kg/m³)	Internal energy (kJ/kg)	Enthalpy (kJ/kg)	Entropy (kJ/kg K)	C_v (kJ/kg K)	C_p (kJ/kg K)	Sound speed (m/s)
130	1	1.8568	1109	1647.6	32.75	7.968	12.246	911.93
140	1	1.7226	1190.8	1771.3	33.666	8.2363	12.49	940.84
150	1	1.6068	1275	1897.3	34.536	8.4853	12.719	968.8
160	1	1.5058	1361.5	2025.6	35.364	8.7133	12.932	995.98
170	1	1.4169	1450.1	2155.9	36.154	8.9199	13.126	1022.5
180	1	1.338	1540.7	2288.1	36.909	9.1056	13.301	1048.4
190	1	1.2675	1632.9	2421.9	37.632	9.2718	13.459	1073.8
200	1	1.2041	1726.7	2557.2	38.326	9.4196	13.599	1098.7
210	1	1.1468	1821.8	2693.8	38.993	9.5506	13.724	1123.2
220	1	1.0947	1918.1	2831.6	39.634	9.6663	13.834	1147.2
230	1	1.0472	2015.5	2970.4	40.251	9.768	13.931	1170.9
240	1	1.0037	2113.8	3110.2	40.846	9.8573	14.017	1194.2
250	1	0.96362	2213	3250.7	41.42	9.9352	14.091	1217.1
260	1	0.92666	2312.8	3392	41.974	10.003	14.156	1239.7
270	1	0.89244	2413.3	3533.8	42.509	10.062	14.212	1262
280	1	0.86066	2514.3	3676.2	43.027	10.113	14.26	1283.9
290	1	0.83108	2615.8	3819	43.528	10.156	14.302	1305.6
300	1	0.80348	2717.6	3962.2	44.013	10.194	14.337	1326.9
400	1	0.60326	3748.2	5405.9	48.165	10.36	14.492	1526.4
500	1	0.48301	4786.3	6856.7	51.403	10.393	14.52	1704.1
600	1	0.40276	5827.3	8310.2	54.053	10.428	14.554	1864.6
700	1	0.34538	6873.1	9768.4	56.3	10.493	14.617	2011.2
800	1	0.30232	7926.8	11.235	58.258	10.588	14.711	2146.5
900	1	0.26881	8991.6	12.712	59.998	10.714	14.837	2272.2
1000	1	0.24198	10.070	14.203	61.569	10.87	14.992	2389.8
2.0 MPa								
14.604	2	78.101	−53.009	−27.401	−3.014	5.2144	6.9568	1319.1
20	2	73.526	−12.165	15.036	−0.5571	5.649	8.8908	1216.6
30	2	60.149	99.606	132.86	4.1274	6.2411	16.184	900.05
40	2	19.391	376.53	479.67	13.882	6.9897	27.398	485.25
50	2	11.56	491.84	664.85	18.06	6.5165	14.743	575.26
60	2	8.8198	573.65	800.41	20.537	6.4554	12.745	643.47
70	2	7.2546	648.02	923.7	22.439	6.5143	12.024	699.53
80	2	6.2067	720.14	1042.4	24.024	6.6629	11.762	747.22
90	2	5.4439	792.3	1159.7	25.405	6.879	11.729	788.84
100	2	4.8587	865.74	1277.4	26.645	7.1391	11.825	826.05
110	2	4.3931	941.17	1396.4	27.78	7.4214	11.993	860.11
120	2	4.0126	1018.9	1517.4	28.832	7.7089	12.198	891.91
130	2	3.695	1099.1	1640.4	29.817	7.9896	12.416	922.06
140	2	3.4255	1181.8	1765.7	30.745	8.2558	12.634	950.95
150	2	3.1937	1266.8	1893.1	31.624	8.5031	12.843	978.86
160	2	2.992	1354	2022.5	32.459	8.7296	13.039	1006
170	2	2.8148	1443.3	2153.8	33.255	8.935	13.22	1032.4
180	2	2.6578	1534.3	2286.8	34.015	9.1198	13.384	1058.2
190	2	2.5176	1627	2421.4	34.743	9.2851	13.532	1083.5

(continued)

Table 4.12 (continued)

Temperature (K)	Pressure (MPa)	Density (kg/m^3)	Internal energy (kJ/kg)	Enthalpy (kJ/kg)	Entropy (kJ/kg K)	C_v (kJ/kg K)	C_p (kJ/kg K)	Sound speed (m/s)
200	2	2.3918	1721.2	2557.4	35.441	9.4322	13.665	1108.3
210	2	2.278	1816.7	2694.7	36.11	9.5625	13.783	1132.7
220	2	2.1748	1913.4	2833	36.754	9.6776	13.887	1156.6
230	2	2.0806	2011.1	2972.4	37.373	9.7789	13.98	1180.2
240	2	1.9942	2109.7	3112.6	37.97	9.8676	14.06	1203.3
250	2	1.9149	2209.1	3253.5	38.545	9.9451	14.131	1226.1
260	2	1.8417	2309.2	3395.2	39.101	10.013	14.192	1248.6
270	2	1.7739	2409.9	3537.4	39.637	10.071	14.246	1270.8
280	2	1.7109	2511.1	3680.1	40.156	10.121	14.291	1292.6
290	2	1.6523	2612.7	3823.2	40.659	10.165	14.33	1314.2
300	2	1.5976	2714.8	3966.6	41.145	10.202	14.364	1335.4
400	2	1.2008	3746.7	5412.2	45.303	10.366	14.505	1533.9
500	2	0.96224	4785.6	6864.1	48.542	10.397	14.528	1710.8
600	2	0.80286	5827.1	8318.2	51.193	10.432	14.558	1870.6
700	2	0.68881	6873.2	9776.8	53.442	10.496	14.62	2016.7
800	2	0.60315	7927.2	11243	55.4	10.591	14.713	2151.6
900	2	0.53644	8992.2	12720	57.14	10.717	14.839	2277
1000	2	0.48303	10071	14212	58.711	10.872	14.993	2394.2
5.0 MPa								
15.518	5	79.609	−50.974	11.833	−2.9347	5.2713	6.9083	1394.6
20	5	76.333	−19.778	45.725	−1.0233	5.6472	8.2607	1326.5
30	5	66.614	71.978	147.04	3.0304	6.2734	12.278	1116.7
40	5	51.723	203.8	300.47	7.3983	6.5792	18.966	844.83
50	5	33.785	362.99	510.99	12.088	6.7136	20.751	694.08
60	5	23.813	485.08	695.05	15.456	6.647	16.374	708.27
70	5	18.774	580.23	846.56	17.796	6.6678	14.22	750.04
80	5	15.705	664.92	983.3	19.623	6.7883	13.252	791.81
90	5	13.597	745.63	1113.4	21.156	6.9841	12.821	830.1
100	5	12.04	825.35	1240.6	22.497	7.2291	12.667	865.07
110	5	10.832	905.62	1367.2	23.703	7.5	12.665	897.45
120	5	9.8626	987.25	1494.2	24.808	7.7786	12.748	927.91
130	5	9.0636	1070.6	1622.3	25.833	8.0523	12.876	956.95
140	5	8.3919	1156	1751.8	26.793	8.3128	13.024	984.89
150	5	7.818	1243.3	1882.8	27.697	8.5554	13.179	1012
160	5	7.3213	1332.4	2015.4	28.552	8.7781	13.331	1038.3
170	5	6.8866	1423.4	2149.4	29.365	8.9802	13.476	1064.1
180	5	6.5026	1515.9	2284.9	30.139	9.1621	13.611	1089.3
190	5	6.1606	1610	2421.6	30.878	9.3249	13.734	1114
200	5	5.854	1705.4	2559.5	31.586	9.4697	13.846	1138.3
210	5	5.5773	1802	2698.5	32.264	9.5982	13.946	1162.1
220	5	5.3262	1899.6	2838.4	32.914	9.7115	14.035	1185.5
230	5	5.0974	1998.2	2979.1	33.54	9.8112	14.114	1208.6
240	5	4.8878	2097.7	3120.6	34.142	9.8985	14.183	1231.3
250	5	4.6951	2197.8	3262.8	34.722	9.9747	14.243	1253.7
260	5	4.5174	2298.6	3405.5	35.282	10.041	14.295	1275.8

(continued)

Table 4.12 (continued)

Temperature (K)	Pressure (MPa)	Density (kg/m^3)	Internal energy (kJ/kg)	Enthalpy (kJ/kg)	Entropy (kJ/kg K)	C_v (kJ/kg K)	C_p (kJ/kg K)	Sound speed (m/s)
270	5	4.3528	2400	3548.6	35.822	10.098	14.34	1297.5
280	5	4.2	2501.8	3692.2	36.345	10.148	14.379	1319
290	5	4.0577	2604	3836.2	36.85	10.19	14.411	1340.2
300	5	3.9249	2706.6	3980.5	37.339	10.227	14.439	1361.1
400	5	2.96	3742.3	5431.5	41.513	10.384	14.544	1556.5
500	5	2.3778	4783.5	6886.3	44.759	10.412	14.55	1731
600	5	1.9876	5826.5	8342.1	47.413	10.444	14.572	1888.9
700	5	1.7076	6873.7	9801.8	49.663	10.506	14.628	2033.4
800	5	1.4968	7928.5	11.269	51.622	10.6	14.719	2166.9
900	5	1.3324	8994.2	12.747	53.362	10.725	14.842	2291.2
1000	5	1.2006	10.074	14.238	54.934	10.879	14.995	2407.5
10.0 MPa								
16.929	10	81.803	−46.527	75.718	−2.8144	5.3582	6.8871	1509.8
20	10	79.926	−27.076	98.04	−1.6045	5.6252	7.6607	1471.7
30	10	72.579	50.733	188.51	2.0253	6.3191	10.48	1326.1
40	10	63.362	150.08	307.9	5.4399	6.6438	13.382	1159.3
50	10	52.836	264.99	454.25	8.6959	6.7406	15.669	1012.9
60	10	42.925	381.98	614.95	11.624	6.7767	16.148	932.19
70	10	35.295	489.65	772.97	14.062	6.8319	15.368	910.09
80	10	29.86	587.23	922.12	16.055	6.9485	14.499	916.81
90	10	25.937	678.37	1063.9	17.726	7.1307	13.91	934.69
100	10	22.954	766.31	1201.2	19.172	7.3614	13.584	956.61
110	10	20.703	853.16	1336.2	20.459	7.6195	13.437	979.96
120	10	18.862	940.16	1470.3	21.626	7.8872	13.404	1003.8
130	10	17.347	1028	1604.5	22.7	8.1516	13.44	1027.8
140	10	16.075	1117.2	1739.2	23.699	8.4041	13.514	1051.7
150	10	14.99	1207.7	1874.9	24.634	8.64	13.608	1075.6
160	10	14.051	1299.7	2011.4	25.516	8.8568	13.711	1099.2
170	10	13.23	1393.2	2149.1	26.35	9.0539	13.814	1122.7
180	10	12.504	1488	2287.7	27.143	9.2313	13.913	1145.9
190	10	11.853	1584	2427.3	27.897	9.3902	14.007	1168.9
200	10	11.279	1681.2	2567.8	28.618	9.5316	14.092	1191.6
210	10	10.756	1779.4	2709.1	29.307	9.6569	14.17	1214
220	10	10.28	1878.6	2851.2	29.968	9.7674	14.239	1236.2
230	10	9.8485	1978.5	2993.9	30.603	9.8646	14.301	1258.1
240	10	9.4519	2079.2	3137.2	31.212	9.9496	14.355	1279.8
250	10	9.087	2180.5	3281	31.799	10.024	14.401	1301.2
260	10	8.7501	2282.3	3425.2	32.365	10.088	14.442	1322.3
270	10	8.4381	2384.7	3569.8	32.911	10.144	14.476	1343.2
280	10	8.1481	2487.4	3714.7	33.438	10.191	14.505	1363.8
290	10	7.8779	2590.5	3859.9	33.947	10.232	14.528	1384.3
300	10	7.6254	2693.8	4005.2	34.44	10.267	14.548	1404.4
400	10	5.7843	3735.5	5464.3	38.637	10.414	14.602	1594
500	10	4.6657	4780.3	6923.6	41.893	10.436	14.584	1764.4
600	10	3.9117	5825.7	8382.1	44.553	10.464	14.593	1919

(continued)

Table 4.12 (continued)

Temperature (K)	Pressure (MPa)	Density (kg/m³)	Internal energy (kJ/kg)	Enthalpy (kJ/kg)	Entropy (kJ/kg K)	C_v (kJ/kg K)	C_p (kJ/kg K)	Sound speed (m/s)
700	10	3.3683	6874.7	9843.6	46.805	10.523	14.642	2060.9
800	10	2.9578	7930.8	11.312	48.766	10.614	14.727	2192.3
900	10	2.6366	8997.5	12.790	50.507	10.737	14.847	2314.7
1000	10	2.3784	10.078	14.282	52.079	10.89	14.998	2429.5
50.0 MPa								
25.812	50	93.226	9.2009	545.53	−2.1032	5.8182	6.9715	2065.6
30	50	91.775	31.515	576.32	−0.99911	6.1757	7.7324	2048.7
40	50	87.918	93.264	661.98	1.4539	6.8362	9.3227	2002.9
50	50	83.788	164.49	761.23	3.6641	7.1947	10.46	1952.5
60	50	79.61	242.01	870.07	5.6462	7.3898	11.263	1901.6
70	50	75.517	323.72	985.83	7.4294	7.5366	11.863	1852.4
80	50	71.59	408.5	1106.9	9.0457	7.6944	12.34	1806.3
90	50	67.879	495.77	1232.4	10.523	7.8818	12.74	1765.1
100	50	64.411	585.26	1361.5	11.883	8.0961	13.082	1729.6
110	50	61.195	676.79	1493.9	13.144	8.3266	13.377	1700.4
120	50	58.226	770.22	1628.9	14.32	8.5618	13.633	1677.4
130	50	55.495	865.41	1766.4	15.42	8.7922	13.854	1660.2
140	50	52.984	962.23	1905.9	16.453	9.0113	14.044	1648.1
150	50	50.676	1060.5	2047.2	17.428	9.2152	14.207	1640.4
160	50	48.553	1160.2	2190	18.35	9.4021	14.346	1636.4
170	50	46.596	1261	2334	19.223	9.5713	14.464	1635.5
180	50	44.79	1362.9	2479.2	20.053	9.7231	14.565	1637.2
190	50	43.118	1465.7	2625.3	20.842	9.8583	14.649	1641.1
200	50	41.569	1569.3	2772.1	21.596	9.978	14.72	1646.7
210	50	40.128	1673.6	2919.6	22.315	10.083	14.779	1653.8
220	50	38.787	1778.6	3067.7	23.004	10.175	14.827	1662
230	50	37.535	1884.1	3216.2	23.664	10.255	14.866	1671.3
240	50	36.363	1990	3365	24.297	10.325	14.897	1681.4
250	50	35.265	2096.2	3514.1	24.906	10.384	14.921	1692.2
260	50	34.233	2202.8	3663.4	25.492	10.435	14.938	1703.6
270	50	33.261	2309.6	3812.8	26.056	10.478	14.95	1715.4
280	50	32.346	2416.6	3962.4	26.599	10.514	14.958	1727.7
290	50	31.481	2523.7	4112	27.124	10.544	14.961	1740.3
300	50	30.662	2630.9	4261.6	27.632	10.568	14.961	1753.1
400	50	24.381	3703	5753.7	31.925	10.64	14.865	1888
500	50	20.276	4768.4	7234.4	35.229	10.616	14.756	2022.6
600	50	17.371	5828.6	8707	37.914	10.613	14.707	2151.1
700	50	15.201	6888.7	10.178	40.182	10.65	14.718	2272.6
800	50	13.517	7953.4	11.652	42.151	10.724	14.778	2387.4
900	50	12.171	9026.8	13.135	43.897	10.834	14.88	2496
1000	50	11.069	10.113	14.630	45.471	10.976	15.018	2599
100.0 MPa								
34.272	100	102.18	92.484	1071.1	−1.6019	6.0767	7.2172	2505.7
40	100	100.64	121.65	1115.3	−0.413	6.6124	8.1628	2494.4
50	100	97.741	180.4	1203.5	1.5515	7.2349	9.4138	2467.3

(continued)

Table 4.12 (continued)

Temperature (K)	Pressure (MPa)	Density (kg/m^3)	Internal energy (kJ/kg)	Enthalpy (kJ/kg)	Entropy (kJ/kg K)	C_v (kJ/kg K)	C_p (kJ/kg K)	Sound speed (m/s)
60	100	94.727	246.63	1302.3	3.3499	7.6075	10.294	2435.9
70	100	91.713	318.33	1408.7	4.9887	7.8725	10.96	2402.2
80	100	88.758	394.44	1521.1	6.489	8.1076	11.508	2367.3
90	100	85.896	474.43	1638.6	7.8726	8.3439	11.987	2332.4
100	100	83.146	557.97	1760.7	9.1581	8.5871	12.416	2298.7
110	100	80.516	644.8	1886.8	10.36	8.8323	12.803	2267.3
120	100	78.009	734.68	2016.6	11.489	9.0723	13.149	2238.9
130	100	75.626	827.33	2149.6	12.554	9.3007	13.455	2213.8
140	100	73.362	922.47	2285.6	13.561	9.5132	13.725	2192.2
150	100	71.215	1019.8	2424	14.516	9.7075	13.958	2173.9
160	100	69.178	1119.1	2564.6	15.424	9.8829	14.16	2158.7
170	100	67.246	1220	2707.1	16.287	10.039	14.333	2146.3
180	100	65.413	1322.5	2851.2	17.111	10.178	14.48	2136.5
190	100	63.674	1426.1	2996.6	17.897	10.3	14.604	2129.1
200	100	62.021	1530.9	3143.2	18.649	10.406	14.709	2123.7
210	100	60.451	1636.5	3290.8	19.369	10.498	14.796	2120.1
220	100	58.957	1742.9	3439.1	20.059	10.577	14.867	2118.1
230	100	57.535	1850	3588.1	20.721	10.645	14.925	2117.5
240	100	56.18	1957.5	3737.6	21.357	10.702	14.972	2118.2
250	100	54.887	2065.5	3887.5	21.969	10.75	15.009	2120.1
260	100	53.653	2173.9	4037.7	22.559	10.79	15.037	2122.9
270	100	52.474	2282.5	4188.2	23.126	10.823	15.057	2126.6
280	100	51.347	2391.3	4338.8	23.674	10.849	15.071	2131.1
290	100	50.268	2500.2	4489.6	24.203	10.869	15.079	2136.2
300	100	49.234	2609.3	4640.4	24.715	10.885	15.082	2142
400	100	40.872	3698.3	6144.9	29.044	10.886	14.983	2221.1
500	100	34.982	4777.5	7636.2	32.372	10.815	14.85	2317.6
600	100	30.599	5848.9	9117	35.072	10.781	14.777	2417.8
700	100	27.205	6918	10.594	37.348	10.793	14.769	2517.1
800	100	24.496	7990.1	12.072	39.323	10.85	14.813	2613.8
900	100	22.28	9069.7	13.558	41.072	10.946	14.903	2707.2
1000	100	20.434	10.161	15.054	42.649	11.076	15.032	2797.2

4.7 Formulation for Orthohydrogen

No experimental measurements have ever been taken on pure orthohydrogen. The formulation presented here further extended the Quantum Law of Corresponding States technique validated to predict normal hydrogen properties to predict full orthohydrogen properties. The original parahydrogen equation of state was then refit to the orthohydrogen surface. Uncertainties are assumed to be similar to those for normal hydrogen (Tables 4.13, 4.14, 4.15 and 4.16).

Table 4.13 Fixed point properties and correlation limits for orthohydrogen

	Temperature (K)	Pressure (kPa)	Density (mol/dm^3)
Critical point	33.22	1310.65	15.445
Triple point	14.008	7.461	
Upper limit	1000	2,000,000	38.2
Lower limit	14.008		

Table 4.14 Fluid constants and normal boiling point for orthohydrogen

Molar mass	2.01588 g/mol
Ideal-gas constant	8.314472 J/(mol K)
Pitzer's acentric factor	−0.219
Normal boiling point	20.380 K

Table 4.15 Fundamental equation of state coefficients and exponents for orthohydrogen

i	N_i	t_i	d_i	p_i	φ_i	β_i	γ_i	D_i
1	−6.83148	0.7333	1					
2	0.01	1	4					
3	2.11505	1.1372	1					
4	4.38353	0.5136	1					
5	0.211292	0.5638	2					
6	−1.00939	1.6248	2					
7	0.142086	1.829	3					
8	−0.87696	2.404	1	1				
9	0.804927	2.105	3	1				
10	−0.710775	4.1	2		1.169	0.4555	1.5444	0.6366
11	0.0639688	7.658	1		0.894	0.4046	0.6627	0.3876
12	0.0710858	1.259	3		0.04	0.0869	0.763	0.9437
13	−0.087654	7.589	1		2.072	0.4415	0.6587	0.3976
14	0.647088	3.946	1		1.306	0.5743	1.4327	0.9626

Table 4.16 Ancillary function coefficients and exponents for orthohydrogen

i	α^0	
	a_i	k_i
0	2.5	
1	−1.4675442336	
2	1.8845068862	
3	2.54151	856.0
4	−2.3661	1444.0
5	1.00365	2194.0
6	1.22447	6968.0

$m = 0, n = 4$

i	Vapor pressure		Saturated liquid density		Saturated vapor density	
	N_i	k_i	N_i	k_i	N_i	k_i
1	−4.8868	1	4.3911	0.53	−3.1463	0.491
2	1.05310	1.5	−7.5872	0.93	−16.183	2.1
3	0.856947	2.7	10.402	1.35	31.803	2.9
4	−0.1854	6.2	−7.2651	1.8	−219.61	4.4
5			1.8302	2.4	431.23	5
6					−255.91	5.5

References

1. C. Gearhart, “Astonishing Successes” and “Bitter Disappointment”: the specific heat of hydrogen in quantum theory. Arch. Hist. Exact Sci. **64**, 113–202 (2010)
2. J.W. Leachman, R.T. Jacobsen, S.G. Penoncello, E.W. Lemmon, Fundamental equations of state for parahydrogen, normal hydrogen, and orthohydrogen. J. Phys. Chem. Ref. Data. **38**(3), 721–748 (2009)

Chapter 5
Deuterium

The previous chapter established the role of quantum mechanics in small changes of the hydrogen molecule that resulted in macroscopic changes to bulk thermophysical properties. In this chapter the role of additional neutrons in hydrogen nuclei is considered. Figure 5.1 shows the isotopic and non-isotopic (isotopologues) of the hydrogen molecule. Deuterium is a stable isotope of hydrogen with an additional neutron that doubles the atomic weight of a nucleus. The ratio of deuterium to hydrogen naturally occurring on Earth is an appreciable 1 in 7200 atoms. Tritium is a non-stable, radioactive form of hydrogen with two additional neutrons in the nucleus. The short half-life of tritium necessitates production of tritium via nuclear processes and the expense potentially exceeds even that of ^{3}He. The non-isotopic molecular combinations of the atoms (hydrogen-deuteride, hydrogen-tritide, and deuterium-tritide) do not have ortho-para spin-isomers. However, the isotopic molecules (hydrogen, deuterium, and tritium) do. A compilation of the available experimental measurements of the hydrogen isotopologues is available [1]. Experimental studies are mostly limited to deuterium and hardly justify wide-ranging formulation development.

Contrary to hydrogen and tritium, deuterium follows Bose-Einstein statistics resulting in naming conventions that are easily confused. Hydrogen and tritium have a 3:1 ratio of ortho to para spin-isomers with the form less prevalent at room temperature (para) occupying the even rotational energy levels ($J = 0, 2, 4\ldots$). However, deuterium has a 3:6 ratio of para to ortho with the form less prevalent at room temperature (para) occupying odd rotational energy levels ($J = 1, 3, 5\ldots$). The resulting equilibrium composition of normal deuterium at room temperature is 66.6% orthodeuterium and 33.3% paradeuterium. Figure 5.2 shows the equilibrium orthodeuterium fraction versus temperature. The equilibrium ortho-para composition change for deuterium is small (33.3%) relative to hydrogen or tritium (75%). Differences in para-ortho heat capacities are more concentrated and reach a peak at approximately half the temperature ($\sim$70 K) of hydrogen. The overall differences in heat capacities remain substantial and three ideal-gas equations for the reduced

J.W. Leachman et al., *Thermodynamic Properties of Cryogenic Fluids*,
International Cryogenics Monograph Series, DOI 10.1007/978-3-319-57835-4_5

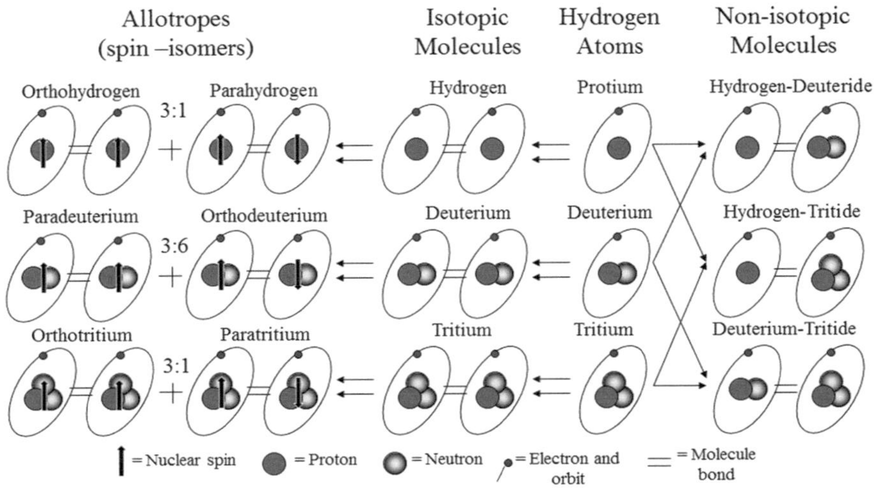

Fig. 5.1 Hydrogen isotopologues and allotropes (nuclear spin-isomers)

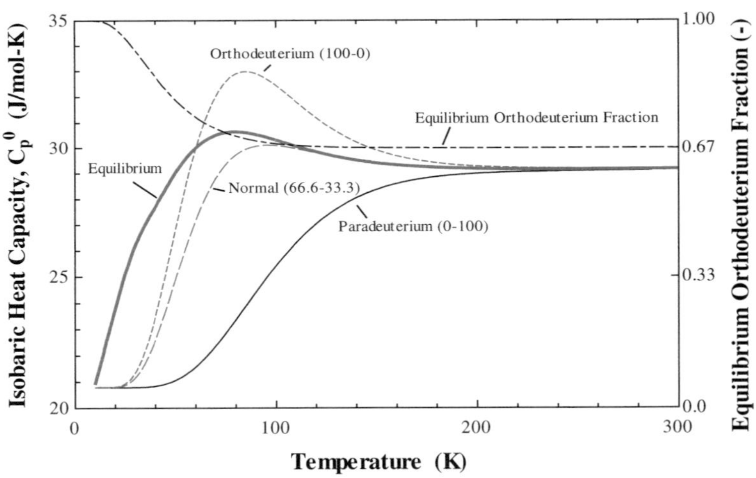

Fig. 5.2 Ideal-gas isobaric heat capacities and equilibrium composition of deuterium versus temperature

Helmholtz energy a^0, are provided. The property tables and plots that follow assume a normal deuterium spin-isomer composition.

The relatively small composition changes between normal and orthodeuterium result in real fluid property differences, such as vapor pressures and liquid phase densities, that are often less than the resolution capabilities of the experiments. Indeed, the few liquid phase measurements that are available seldom specified para-ortho compositions for deuterium and are often assumed normal. As a result, only a normal deuterium real-fluid residual a^r, was developed for the formulation.

5.1 Formulation for Deuterium

A fundamental equation of state for deuterium was developed by Richardson et al. [2]. The uncertainty in predicted density is 0.5% over the valid temperature range and pressures up to 300 MPa. The uncertainties of vapor pressures and saturated liquid densities are 2 and 3%, respectively, while speed-of-sound values are accurate to within 1% in the liquid phase (Tables 5.1, 5.2, 5.3 and 5.4).

Table 5.1 Fixed point properties and correlation limits for deuterium

	Temperature (K)	Pressure (kPa)	Density (mol/dm^3)
Critical point	38.34	1679.6	17.23
Triple point	18.71		
Upper limit	600	2,000,000	96.7
Lower limit	18.71		

Table 5.2 Fluid constants and normal boiling point for deuterium

Molar mass	4.0282 g/mol
Ideal gas constant	8.3144621 J/(mol K)
Pitzer's acentric factor	−0.175
Normal boiling point	23.647 K

Table 5.3 Fundamental equation of state coefficients and exponents for deuterium

i	N_i	t_i	d_i	p_i	φ_i	β_i	γ_i	D_i
1	0.006267958	1	4					
2	10.53609	0.462	1					
3	−10.14149	0.5584	1					
4	0.356061	0.627	2					
5	0.1824472	1.201	3					
6	−1.129638	0.309	1					
7	−0.0549812	1.314	3					
8	−0.6791329	1.1166	2					
9	1.347918	1.25	2	1				
10	−0.8657582	1.25	2	1				
11	1.719146	1.395	1	2				
12	−1.917977	1.627	1	2				
13	0.1233365	1	3	2				
14	−0.07936891	2.5	2	2				
15	1.686617	0.635	1		0.868	0.613	0.6306	1.46

(continued)

Table 5.3 (continued)

i	N_i	t_i	d_i	p_i	φ_i	β_i	γ_i	D_i
16	−4.240326	0.664	1		0.636	0.584	0.711	1.7864
17	1.857114	0.7082	2		0.668	0.57	0.6446	1.647
18	−0.5903705	2.25	3		0.65	1.056	0.8226	0.541
19	1.520171	1.524	3		0.745	1.01	0.992	0.969
20	2.361373	0.67	1		0.782	1.025	1.2184	1.892
21	−2.297315	0.709	3		0.693	1.029	1.203	1.076

Table 5.4 Ancillary function coefficients and exponents for deuterium

	α^0					
	Normal deuterium		Orthodeuterium		Paradeuterium	
i	a_i		a_i		a_i	
1	−2.0677351753		−2.0672670563		−2.0683998716	
2	2.4237151502		2.4234599781		2.4241000701	
i	a_i	k_i	a_i	k_i	a_i	k_i
0	2.5		2.5		2.5	
3	−3.54145	7174.1	4.04482	1591	1.28527	5068
4	3.03260	8635	−4.65391	481.6	1.11376	1000.8
5	−3.52422	902.7	−4.65342	472.4	−2.49100	261.5
6	−1.73421	181.1	3.46313	362.2	6.38763	437.2
7	−3.57135	438.5	−4.58637	2038	6.17406	312.3
8	2.14858	5034.2	−4.65030	463.2	−3.13698	382.8
9	6.23107	269.9	−4.65124	491.3	−3.14254	356.8
10	−3.30425	229.9	2.67024	2713.4	−2.29511	294.7
11	6.23098	666.4	15.20455	618.6	−3.37000	682.4
12	−3.57137	452.8	0.87164	8642	1.13634	246
13	3.32901	192	−4.76080	961.7	0.72512	277.1
14	0.97782	1187.6	4.32447	253.2	–	–

$n = 14$ (there are no polynomial terms in the first summation)

	Vapor pressure		Saturated liquid density		Saturated vapor density	
i	N_i	k_i	N_i	k_i	N_i	k_i
1	−5.5706	1	3.3769	0.512	−3.8111	0.528
2	1.7631	1.5	−5.3693	1.12	−7.3624	2.03
3	−0.5458	2.83	11.943	1.8	2.2294	3.6
4	1.2154	4.06	−17.361	2.55	−21.443	5.0
5	−1.1556	5.4	15.170	3.4	12.796	6.5
6			−6.3079	4.4	−31.334	9.0

5.2 Property Plots for Deuterium

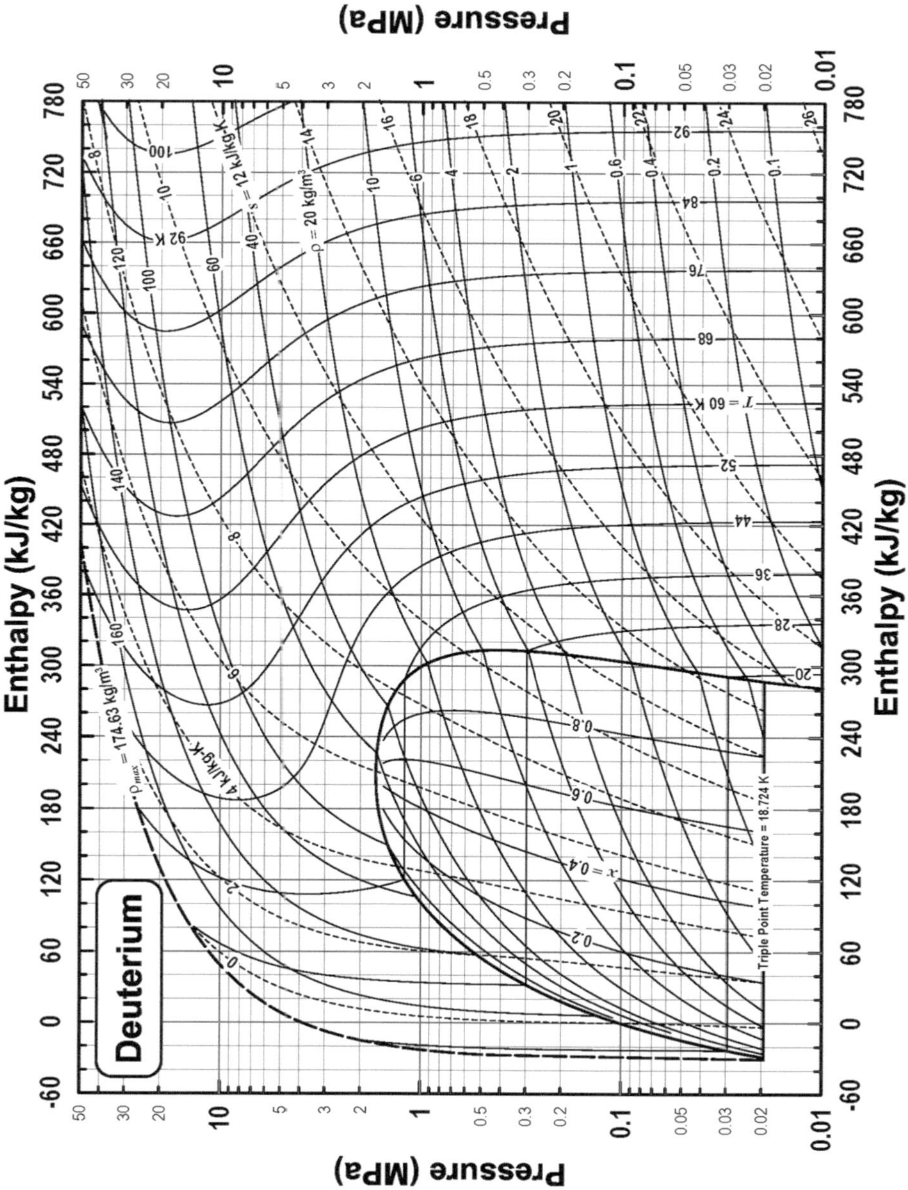
Deuterium
Enthalpy (kJ/kg)
Pressure (MPa)
Triple Point Temperature = 18.724 K
ρ_{max} = 174.63 kg/m³
ρ = 20 kg/m³
s = 12 kJ/kg-K
4 kJ/kg-K
T = 60 K
92 K
x = 0.4

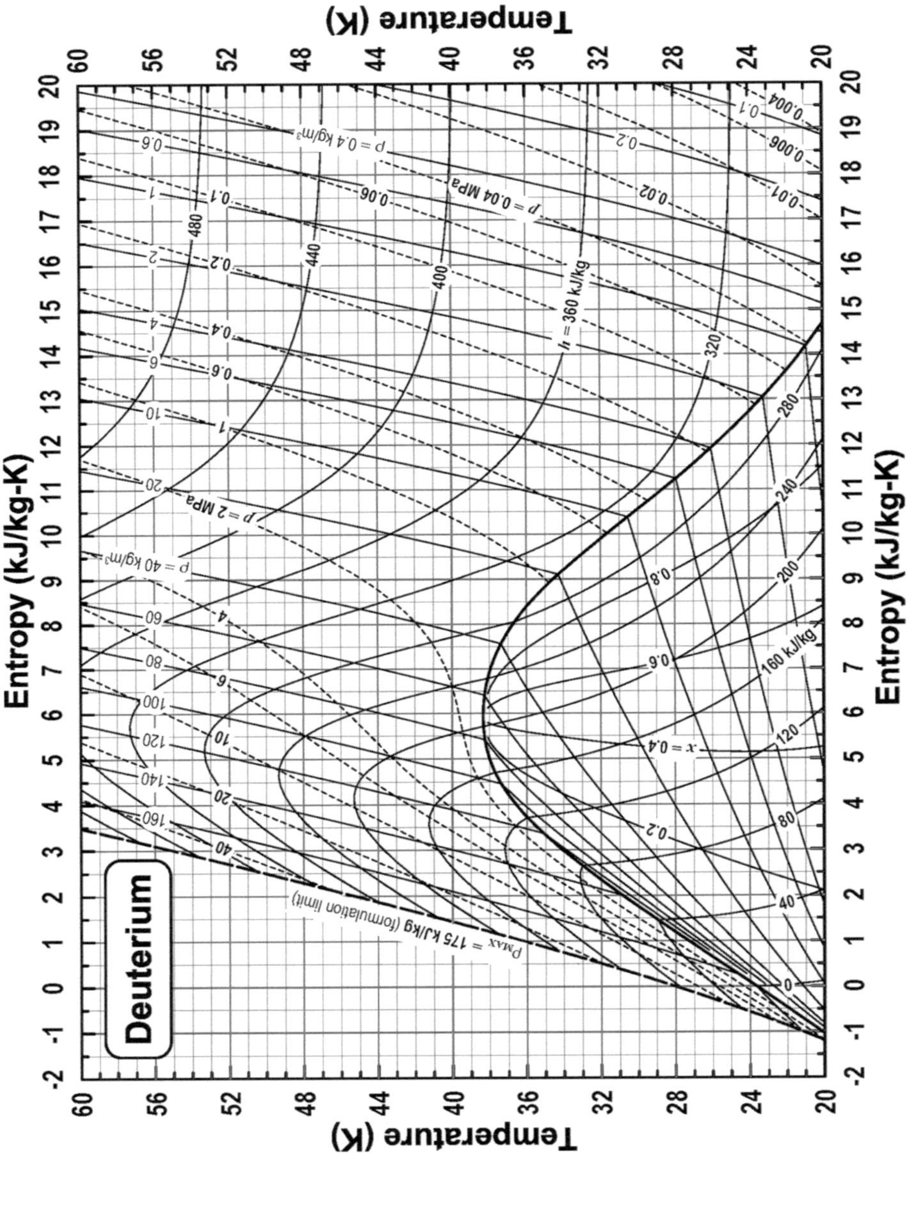
Deuterium
Entropy (kJ/kg-K)
Temperature (K)
p = 2 MPa
p = 0.04 MPa
h = 360 kJ/kg
160 kJ/kg
x = 0.4

5.3 Thermodynamic Data Tables for Deuterium

See Tables 5.5 and 5.6.

Table 5.5 Thermodynamic properties of saturated deuterium

Temperature (K)	Pressure (MPa)	Liquid density (kg/m^3)	Vapor density (kg/m^3)	Liquid enthalpy (kJ/kg)	Vapor enthalpy (kJ/kg)	Liquid entropy (kJ/kg K)	Vapor entropy (kJ/kg K)	Liquid C_v (kJ/kg K)	Vapor C_v (kJ/kg K)	Liquid sound speed (m/s)	Vapor sound speed (m/s)
18.71	0.017189	174.31	0.45536	−30.271	285.81	−1.4075	15.486	3.3985	3.1452	5.5683	5.3651
19	0.019541	173.67	0.51086	−28.639	287.06	−1.3216	15.294	3.4026	3.1498	5.6191	5.3852
20	0.029587	171.4	0.74116	−22.885	291.2	−1.0295	14.675	3.4171	3.1678	5.8047	5.4649
21	0.043147	169.06	1.0399	−16.924	295.09	−0.74256	14.115	3.4321	3.1891	6.0069	5.5622
22	0.06092	166.64	1.4184	−10.736	298.69	−0.45967	13.605	3.4474	3.2137	6.2264	5.6796
23	0.083638	164.14	1.8886	−4.3026	301.97	−0.17983	13.136	3.4629	3.2416	6.4646	5.8199
23.647	0.10133	162.47	2.2477	0	303.91	0	12.852	3.4731	3.2615	6.6295	5.9247
24	0.11206	161.54	2.4635	2.3971	304.9	0.097838	12.702	3.4788	3.273	6.7235	5.987
25	0.14696	158.83	3.1575	9.385	307.44	0.37417	12.296	3.4957	3.3079	7.0067	6.1859
26	0.18914	156.01	3.9866	16.686	309.56	0.64997	11.914	3.514	3.3466	7.3191	6.4234
27	0.2394	153.06	4.9697	24.329	311.21	0.92611	11.551	3.5344	3.3894	7.6684	6.7086
28	0.29857	149.96	6.1288	32.349	312.34	1.2035	11.203	3.5576	3.4367	8.0655	7.0543
29	0.36748	146.69	7.4911	40.79	312.89	1.4834	10.866	3.5844	3.489	8.5263	7.4788
30	0.447	143.22	9.0905	49.708	312.78	1.7671	10.536	3.6156	3.5469	9.0747	8.009
31	0.53803	139.5	10.972	59.176	311.91	2.0564	10.209	3.652	3.6109	9.7479	8.6868
32	0.6415	135.49	13.195	69.298	310.14	2.3538	9.8801	3.6946	3.6818	10.607	9.5806
33	0.75844	131.1	15.849	80.221	307.27	2.6628	9.5432	3.7445	3.7605	11.761	10.811
34	0.88999	126.2	19.066	92.173	303.01	2.9891	9.1902	3.8034	3.8483	13.42	12.612
35	1.0375	120.55	23.073	105.54	296.85	3.3417	8.8078	3.8739	3.947	16.057	15.507
36	1.2026	113.74	28.303	121.06	287.85	3.7391	8.3721	3.9608	4.0595	20.993	20.94
37	1.3875	104.67	35.81	140.56	273.76	4.2266	7.8268	4.0746	4.191	33.726	34.835
38	1.5962	88.684	50.168	172.13	245.38	5.0102	6.9379	4.2538	4.3503	130.24	136.54
38.34	1.6738	69.406	69.406	208.12	208.12	5.9268	5.9268				

Table 5.6 Thermodynamic properties of deuterium

Temperature (K)	Pressure (MPa)	Density (kg/m^3)	Internal energy (kJ/kg)	Enthalpy (kJ/kg)	Entropy (kJ/kg K)	C_v (kJ/kg K)	C_p (kJ/kg K)	Sound speed (m/s)
0.1 MPa								
18.74	0.1	174.36	−30.317	−29.743	−1.4046	3.399	5.5673	1078.7
20	0.1	171.51	−23.169	−22.586	−1.035	3.4169	5.798	1053.4
23.601	0.1	162.59	−0.9211	−0.30604	−0.01261	3.4724	6.6177	971.33
23.601	0.1	2.221	258.75	303.78	12.871	3.2601	5.9169	273.95
30	0.1	1.6806	280.83	340.33	14.243	3.2516	5.599	313.55
40	0.1	1.2331	315.62	396.72	15.864	3.5696	5.7667	358.68
50	0.1	0.97811	354.54	456.78	17.202	4.1303	6.272	392.17
60	0.1	0.81175	399.04	522.23	18.394	4.6889	6.8037	421.67
70	0.1	0.69424	448.31	592.36	19.475	5.0925	7.192	450.34
80	0.1	0.60667	500.65	665.48	20.451	5.3167	7.4069	478.79
90	0.1	0.53883	554.46	740.05	21.329	5.4046	7.4886	506.86
100	0.1	0.48469	608.68	814.99	22.118	5.4114	7.4911	534.31
110	0.1	0.44047	662.72	889.75	22.831	5.3802	7.4568	560.92
120	0.1	0.40367	716.37	964.09	23.478	5.3374	7.4117	586.56
130	0.1	0.37256	769.58	1038	24.069	5.2965	7.3691	611.2
140	0.1	0.34591	822.41	1111.5	24.614	5.2629	7.3341	634.85
150	0.1	0.32283	874.94	1184.7	25.119	5.2375	7.3076	657.58
160	0.1	0.30263	927.24	1257.7	25.59	5.2192	7.2885	679.48
170	0.1	0.28482	979.39	1330.5	26.032	5.2067	7.2752	700.62
180	0.1	0.26899	1031.4	1403.2	26.447	5.1982	7.2662	721.09
190	0.1	0.25483	1083.4	1475.8	26.84	5.1926	7.2601	740.95
200	0.1	0.24209	1135.3	1548.4	27.212	5.1889	7.256	760.26
210	0.1	0.23056	1187.2	1620.9	27.566	5.1865	7.2532	779.07
220	0.1	0.22008	1239.1	1693.5	27.904	5.1848	7.2512	797.42
230	0.1	0.21051	1290.9	1766	28.226	5.1837	7.2498	815.35
240	0.1	0.20174	1342.8	1838.5	28.534	5.1829	7.2489	832.89
250	0.1	0.19367	1394.6	1911	28.83	5.1824	7.2482	850.05
260	0.1	0.18623	1446.4	1983.4	29.115	5.1822	7.2477	866.88
270	0.1	0.17933	1498.3	2055.9	29.388	5.1821	7.2475	883.37
280	0.1	0.17293	1550.1	2128.4	29.652	5.1822	7.2475	899.56
290	0.1	0.16697	1601.9	2200.9	29.906	5.1825	7.2476	915.46
300	0.1	0.1614	1653.8	2273.3	30.152	5.1829	7.248	931.09
400	0.1	0.12106	2172.6	2998.6	32.238	5.1963	7.2608	1074.6
500	0.1	0.096859	2693.7	3726.1	33.861	5.2288	7.293	1200.2
600	0.1	0.080721	3219.4	4458.2	35.196	5.2908	7.3549	1312.5
0.2 MPa								
20	0.2	171.66	−23.325	−22.16	−1.0429	3.4168	5.7886	1056.2
26.231	0.2	155.34	17.132	18.419	0.71367	3.5185	7.3962	901.07
26.231	0.2	4.1992	262.36	309.99	11.829	3.3561	6.4846	282.1
30	0.2	3.5155	276.56	333.45	12.666	3.3204	6.0385	307.29
40	0.2	2.5129	312.97	392.56	14.367	3.5915	5.9367	355.96
50	0.2	1.975	352.6	453.87	15.733	4.1411	6.3647	390.77

(continued)

Table 5.6 (continued)

Temperature (K)	Pressure (MPa)	Density (kg/m^3)	Internal energy (kJ/kg)	Enthalpy (kJ/kg)	Entropy (kJ/kg K)	C_v (kJ/kg K)	C_p (kJ/kg K)	Sound speed (m/s)
60	0.2	1.6321	397.52	520.06	16.938	4.6956	6.8626	420.95
70	0.2	1.3927	447.07	590.68	18.026	5.0972	7.2329	450.04
80	0.2	1.2155	499.61	664.15	19.007	5.3203	7.4368	478.76
90	0.2	1.0787	553.56	738.98	19.888	5.4075	7.5115	507.03
100	0.2	0.96978	607.9	814.13	20.68	5.4138	7.5091	534.62
110	0.2	0.88101	662.03	889.05	21.394	5.3821	7.4713	561.32
120	0.2	0.80722	715.76	963.52	22.042	5.3391	7.4236	587.04
130	0.2	0.74488	769.03	1037.5	22.635	5.298	7.379	611.72
140	0.2	0.69152	821.91	1111.1	23.18	5.2642	7.3425	635.41
150	0.2	0.64532	874.49	1184.4	23.686	5.2386	7.3148	658.17
160	0.2	0.60493	926.83	1257.4	24.157	5.2203	7.2947	680.08
170	0.2	0.5693	979.01	1330.3	24.599	5.2077	7.2806	701.24
180	0.2	0.53765	1031.1	1403.1	25.015	5.1991	7.2709	721.71
190	0.2	0.50934	1083.1	1475.8	25.408	5.1935	7.2643	741.57
200	0.2	0.48387	1135	1548.4	25.78	5.1897	7.2597	760.88
210	0.2	0.46082	1186.9	1620.9	26.134	5.1872	7.2565	779.69
220	0.2	0.43988	1238.8	1693.5	26.472	5.1855	7.2542	798.04
230	0.2	0.42076	1290.7	1766	26.794	5.1843	7.2526	815.97
240	0.2	0.40323	1342.6	1838.6	27.103	5.1835	7.2513	833.5
250	0.2	0.38711	1394.4	1911.1	27.399	5.183	7.2504	850.66
260	0.2	0.37223	1446.3	1983.6	27.683	5.1827	7.2498	867.48
270	0.2	0.35845	1498.1	2056.1	27.957	5.1826	7.2494	883.97
280	0.2	0.34565	1549.9	2128.6	28.22	5.1827	7.2492	900.16
290	0.2	0.33374	1601.8	2201	28.475	5.1829	7.2492	916.05
300	0.2	0.32262	1653.6	2273.5	28.72	5.1834	7.2494	931.67
400	0.2	0.24202	2172.5	2998.9	30.807	5.1967	7.2615	1075.1
500	0.2	0.19364	2693.7	3726.5	32.431	5.229	7.2934	1200.7
600	0.2	0.16139	3219.4	4458.6	33.765	5.2911	7.3552	1312.9
0.5 MPa								
20	0.5	172.11	−23.784	−20.879	−1.0661	3.4162	5.7613	1064.5
30.598	0.5	141.03	51.753	55.299	1.9393	3.6367	9.4591	754.6
30.598	0.5	10.178	263.24	312.36	10.341	3.5844	8.3931	288.94
30	0.5	143.43	46.288	49.774	1.757	3.6144	9.0274	780.66
40	0.5	6.6833	304.37	379.18	12.255	3.665	6.5753	347.51
50	0.5	5.0862	346.59	444.9	13.721	4.1747	6.6722	386.68
60	0.5	4.1464	392.89	513.48	14.97	4.7161	7.0487	418.94
70	0.5	3.5136	443.32	585.62	16.081	5.1116	7.3588	449.26
80	0.5	3.0542	496.46	660.17	17.077	5.3311	7.528	478.8
90	0.5	2.7039	550.88	735.79	17.967	5.416	7.5804	507.63
100	0.5	2.4273	605.56	811.55	18.765	5.4208	7.5631	535.61
110	0.5	2.2028	659.97	886.95	19.484	5.388	7.5146	562.6
120	0.5	2.017	713.92	961.82	20.136	5.3441	7.4591	588.51
130	0.5	1.8603	767.38	1036.2	20.731	5.3024	7.4086	613.33

(continued)

Table 5.6 (continued)

Temperature (K)	Pressure (MPa)	Density (kg/m³)	Internal energy (kJ/kg)	Enthalpy (kJ/kg)	Entropy (kJ/kg K)	C_v (kJ/kg K)	C_p (kJ/kg K)	Sound speed (m/s)
140	0.5	1.7265	820.42	1110	21.278	5.2681	7.3675	637.12
150	0.5	1.6108	873.13	1183.5	21.785	5.2422	7.3362	659.95
160	0.5	1.5098	925.6	1256.8	22.258	5.2235	7.3132	681.91
170	0.5	1.4207	977.88	1329.8	22.701	5.2106	7.2968	703.09
180	0.5	1.3416	1030	1402.7	23.118	5.2018	7.2851	723.58
190	0.5	1.2709	1082.1	1475.5	23.511	5.196	7.2768	743.45
200	0.5	1.2073	1134.1	1548.3	23.884	5.192	7.2709	762.77
210	0.5	1.1498	1186.1	1621	24.239	5.1894	7.2665	781.57
220	0.5	1.0976	1238.1	1693.6	24.577	5.1876	7.2632	799.91
230	0.5	1.0499	1290	1766.2	24.9	5.1863	7.2607	817.83
240	0.5	1.0062	1341.9	1838.8	25.209	5.1854	7.2587	835.35
250	0.5	0.96598	1393.8	1911.4	25.505	5.1848	7.2571	852.49
260	0.5	0.92887	1445.7	1984	25.79	5.1844	7.2559	869.3
270	0.5	0.89451	1497.6	2056.5	26.063	5.1842	7.255	885.77
280	0.5	0.86261	1549.4	2129.1	26.327	5.1842	7.2543	901.94
290	0.5	0.8329	1601.3	2201.6	26.582	5.1844	7.2539	917.81
300	0.5	0.80519	1653.2	2274.1	26.828	5.1848	7.2538	933.41
400	0.5	0.6042	2172.3	2999.9	28.915	5.1977	7.2637	1076.6
500	0.5	0.48355	2693.6	3727.6	30.539	5.2298	7.2946	1202.1
600	0.5	0.40308	3219.4	4459.8	31.874	5.2917	7.3559	1314.2
1.0 MPa								
30	1	145.36	43.666	50.546	1.6673	3.6042	8.6449	811.88
34.756	1	122.01	93.926	102.12	3.2526	3.8555	15.279	570.97
34.756	1	22.006	253.12	298.56	8.9046	3.9219	14.651	287.85
40	1	15.239	287.03	352.65	10.363	3.8217	8.4392	332.27
50	1	10.729	335.83	429.04	12.072	4.2351	7.3077	380.34
60	1	8.5208	384.91	502.27	13.407	4.751	7.3927	416.11
70	1	7.1322	436.96	577.17	14.561	5.1355	7.5801	448.41
80	1	6.1582	491.19	653.58	15.581	5.349	7.6839	479.21
90	1	5.4303	546.39	730.54	16.487	5.4302	7.6967	508.91
100	1	4.8625	601.67	807.32	17.296	5.4323	7.6532	537.49
110	1	4.4057	656.55	883.53	18.023	5.3978	7.5866	564.9
120	1	4.0295	710.88	959.05	18.68	5.3525	7.5179	591.1
130	1	3.7139	764.66	1033.9	19.279	5.3098	7.4575	616.14
140	1	3.445	817.96	1108.2	19.83	5.2746	7.4087	640.07
150	1	3.213	870.9	1182.1	20.34	5.248	7.3714	662.99
160	1	3.0107	923.55	1255.7	20.815	5.2288	7.3436	685.02
170	1	2.8327	976.01	1329	21.259	5.2154	7.3233	706.24
180	1	2.6748	1028.3	1402.2	21.677	5.2063	7.3084	726.75
190	1	2.5337	1080.5	1475.2	22.072	5.2001	7.2974	746.63
200	1	2.4069	1132.7	1548.1	22.446	5.1959	7.2892	765.94
210	1	2.2923	1184.7	1621	22.802	5.193	7.2829	784.73
220	1	2.1881	1236.8	1693.8	23.14	5.191	7.2779	803.06

(continued)

Table 5.6 (continued)

Temperature (K)	Pressure (MPa)	Density (kg/m³)	Internal energy (kJ/kg)	Enthalpy (kJ/kg)	Entropy (kJ/kg K)	C_v (kJ/kg K)	C_p (kJ/kg K)	Sound speed (m/s)
230	1	2.0931	1288.8	1766.6	23.464	5.1895	7.274	820.95
240	1	2.0061	1340.8	1839.3	23.773	5.1884	7.2707	838.44
250	1	1.926	1392.8	1912	24.07	5.1877	7.2681	855.56
260	1	1.8521	1444.7	1984.6	24.355	5.1872	7.2659	872.33
270	1	1.7837	1496.6	2057.3	24.629	5.1869	7.2642	888.78
280	1	1.7201	1548.6	2129.9	24.894	5.1868	7.2628	904.91
290	1	1.661	1600.5	2202.6	25.148	5.1868	7.2617	920.75
300	1	1.6058	1652.4	2275.2	25.395	5.1871	7.261	936.32
400	1	1.2056	2171.9	3001.4	27.484	5.1993	7.2673	1079.2
500	1	0.96526	2693.4	3729.4	29.108	5.2311	7.2966	1204.4
600	1	0.80487	3219.3	4461.7	30.443	5.2927	7.3571	1316.3
2.0 MPa								
30	2	148.66	39.223	52.676	1.5116	3.5908	8.107	865.63
40	2	60.758	204.8	237.72	6.5549	4.3672	58.547	323.56
50	2	24.267	310.79	393.21	10.129	4.3708	9.274	371.46
60	2	18.006	367.95	479.02	11.696	4.8219	8.2188	413.03
70	2	14.673	423.91	560.21	12.948	5.1829	8.0628	448.53
80	2	12.498	480.55	640.57	14.021	5.3843	8.0083	481.4
90	2	10.936	537.41	720.3	14.96	5.458	7.9323	512.49
100	2	9.7462	593.93	799.13	15.791	5.4552	7.8331	542.03
110	2	8.804	649.77	876.94	16.532	5.417	7.7287	570.11
120	2	8.0362	704.87	953.74	17.201	5.3691	7.6332	596.79
130	2	7.3963	759.27	1029.7	17.808	5.3243	7.553	622.14
140	2	6.8551	813.1	1104.9	18.366	5.2875	7.4891	646.29
150	2	6.3897	866.48	1179.5	18.88	5.2596	7.4399	669.35
160	2	5.985	919.53	1253.7	19.359	5.2394	7.4027	691.45
170	2	5.6297	972.32	1327.6	19.807	5.2251	7.3748	712.72
180	2	5.3151	1024.9	1401.2	20.228	5.2152	7.3536	733.24
190	2	5.0344	1077.4	1474.7	20.625	5.2084	7.3374	753.11
200	2	4.7823	1129.8	1548	21.001	5.2036	7.3248	772.39
210	2	4.5547	1182.1	1621.2	21.359	5.2002	7.3147	791.15
220	2	4.348	1234.3	1694.3	21.699	5.1978	7.3066	809.42
230	2	4.1595	1286.5	1767.3	22.023	5.1959	7.2999	827.26
240	2	3.9868	1338.6	1840.3	22.334	5.1945	7.2942	844.69
250	2	3.828	1390.7	1913.2	22.631	5.1935	7.2895	861.75
260	2	3.6815	1442.8	1986.1	22.917	5.1927	7.2855	878.45
270	2	3.5459	1494.9	2058.9	23.192	5.1921	7.2821	894.83
280	2	3.42	1546.9	2131.7	23.457	5.1918	7.2792	910.89
290	2	3.3028	1598.9	2204.5	23.712	5.1917	7.2769	926.67
300	2	3.1934	1651	2277.2	23.959	5.1917	7.275	942.16
400	2	2.4001	2171.2	3004.5	26.051	5.2027	7.2743	1084.4
500	2	1.9232	2693	3733	27.677	5.2337	7.3005	1209
600	2	1.6046	3219.2	4465.7	29.012	5.2948	7.3594	1320.5

(continued)

Table 5.6 (continued)

Temperature (K)	Pressure (MPa)	Density (kg/m³)	Internal energy (kJ/kg)	Enthalpy (kJ/kg)	Entropy (kJ/kg K)	C_v (kJ/kg K)	C_p (kJ/kg K)	Sound speed (m/s)
5.0 MPa								
30	5	156.09	29.526	61.558	1.1523	3.5798	7.2547	986.78
40	5	125.84	109.01	148.74	3.6373	4.0195	10.725	698.32
50	5	78.664	221.68	285.24	6.6632	4.615	14.649	458.04
60	5	50.957	312.75	410.87	8.9641	5.003	10.93	444.03
70	5	38.916	383.67	512.15	10.528	5.3085	9.5556	469.77
80	5	32.089	448.64	604.46	11.762	5.4799	8.9688	500.86
90	5	27.573	510.92	692.26	12.796	5.5349	8.6118	532.03
100	5	24.308	571.29	776.98	13.689	5.5192	8.3436	562.04
110	5	21.81	630.05	859.3	14.474	5.4717	8.1284	590.56
120	5	19.821	687.43	939.69	15.173	5.4166	7.9555	617.55
130	5	18.193	743.7	1018.5	15.804	5.3662	7.8189	643.09
140	5	16.829	799.07	1096.2	16.38	5.325	7.7124	667.3
150	5	15.669	853.75	1172.9	16.909	5.2935	7.6302	690.34
160	5	14.666	907.91	1248.8	17.399	5.2702	7.5669	712.35
170	5	13.79	961.68	1324.2	17.857	5.2534	7.5177	733.47
180	5	13.018	1015.1	1399.2	18.285	5.2414	7.4792	753.82
190	5	12.33	1068.4	1473.9	18.689	5.2327	7.4486	773.48
200	5	11.715	1121.4	1548.2	19.07	5.2264	7.4239	792.54
210	5	11.159	1174.3	1622.4	19.432	5.2216	7.4036	811.06
220	5	10.656	1227.1	1696.3	19.776	5.2179	7.3866	829.09
230	5	10.197	1279.7	1770.1	20.104	5.2149	7.3723	846.68
240	5	9.7769	1332.3	1843.8	20.417	5.2126	7.36	863.86
250	5	9.3908	1384.9	1917.3	20.718	5.2106	7.3495	880.67
260	5	9.0348	1437.3	1990.7	21.006	5.209	7.3403	897.13
270	5	8.7052	1489.7	2064.1	21.283	5.2077	7.3324	913.25
280	5	8.3993	1542.1	2137.4	21.549	5.2067	7.3256	929.08
290	5	8.1145	1594.4	2210.6	21.806	5.206	7.3197	944.61
300	5	7.8487	1646.7	2283.8	22.054	5.2055	7.3145	959.87
400	5	5.918	2169	3013.8	24.154	5.2125	7.2942	1100
500	5	4.7536	2692	3743.9	25.783	5.2413	7.3118	1222.9
600	5	3.9732	3219	4477.4	27.121	5.301	7.3662	1333
10.0 MPa								
30	10	164.7	19.085	79.8	0.72268	3.5996	6.628	1124.9
40	10	143.26	84.385	154.19	2.8512	4.0493	8.3067	919
50	10	118.57	162.66	246.99	4.9139	4.6191	10.241	720.36
60	10	93.535	248.13	355.04	6.8817	5.1161	11.025	602.8
70	10	74.675	328.84	462.75	8.5432	5.4367	10.441	571.63
80	10	62.062	402.74	563.87	9.8944	5.5951	9.8091	576.47
90	10	53.32	471.86	659.41	11.02	5.6357	9.3181	593.99
100	10	46.928	537.48	750.57	11.981	5.6079	8.928	616.22
110	10	42.04	600.35	838.22	12.817	5.5504	8.6143	639.93
120	10	38.166	661.05	923.06	13.555	5.487	8.3637	663.71

(continued)

Table 5.6 (continued)

Temperature (K)	Pressure (MPa)	Density (kg/m^3)	Internal energy (kJ/kg)	Enthalpy (kJ/kg)	Entropy (kJ/kg K)	C_v (kJ/kg K)	C_p (kJ/kg K)	Sound speed (m/s)
130	10	35.01	720.04	1005.7	14.216	5.4297	8.1655	686.93
140	10	32.381	777.69	1086.5	14.816	5.3827	8.0099	709.36
150	10	30.151	834.31	1166	15.364	5.3462	7.888	730.97
160	10	28.231	890.14	1244.4	15.87	5.3187	7.7921	751.77
170	10	26.553	945.36	1321.9	16.34	5.2983	7.7161	771.83
180	10	25.085	1000.1	1398.7	16.779	5.2831	7.6551	791.23
190	10	23.777	1054.5	1475	17.192	5.2717	7.6056	810.04
200	10	22.606	1108.5	1550.9	17.581	5.2629	7.5647	828.31
210	10	21.551	1162.3	1626.3	17.949	5.256	7.5305	846.1
220	10	20.594	1215.9	1701.5	18.299	5.2504	7.5016	863.45
230	10	19.723	1269.4	1776.4	18.631	5.2457	7.4768	880.39
240	10	18.925	1322.6	1851	18.949	5.2418	7.4554	896.97
250	10	18.192	1375.8	1925.5	19.253	5.2384	7.4368	913.19
260	10	17.515	1428.8	1999.8	19.545	5.2356	7.4206	929.09
270	10	16.888	1481.8	2073.9	19.824	5.2331	7.4064	944.69
280	10	16.306	1534.7	2147.9	20.093	5.2311	7.3939	960.01
290	10	15.764	1587.5	2221.8	20.353	5.2294	7.3828	975.05
300	10	15.258	1640.2	2295.6	20.603	5.228	7.3732	989.84
400	10	11.57	2165.6	3029.9	22.716	5.2287	7.3245	1126.1
500	10	9.3309	2690.5	3762.2	24.35	5.2539	7.3292	1246
600	10	7.8223	3218.7	4497.1	25.689	5.3113	7.3768	1353.8
50.0 MPa								
30	20	176.45	6.9339	120.28	0.12162	3.681	6.1047	1304.9
40	20	160.57	62.607	187.16	2.0377	4.2107	7.262	1148.7
50	20	144.49	126.92	265.34	3.7776	4.7939	8.355	1003.1
60	20	128.68	198.09	353.52	5.3828	5.2934	9.2223	889.71
70	20	113.91	272.76	448.35	6.8438	5.6159	9.6668	815.15
80	20	100.99	347.41	545.46	8.1404	5.7641	9.6995	775.82
90	20	90.213	419.87	641.57	9.2727	5.7894	9.4989	761.24
100	20	81.382	489.43	735.19	10.259	5.747	9.2215	761.2
110	20	74.122	556.18	826.01	11.125	5.677	8.9456	769.02
120	20	68.087	620.46	914.2	11.893	5.6031	8.6999	781.01
130	20	63.006	682.7	1000.1	12.581	5.5367	8.4917	795.17
140	20	58.672	743.28	1084.2	13.203	5.4819	8.3191	810.44
150	20	54.932	802.53	1166.6	13.772	5.4385	8.1774	826.24
160	20	51.671	860.73	1247.8	14.296	5.405	8.061	842.24
170	20	48.801	918.08	1327.9	14.782	5.3791	7.9649	858.26
180	20	46.255	974.75	1407.1	15.235	5.3591	7.885	874.21
190	20	43.978	1030.9	1485.6	15.659	5.3433	7.8177	890.02
200	20	41.928	1086.5	1563.5	16.059	5.3306	7.7605	905.66
210	20	40.073	1141.8	1640.9	16.436	5.3201	7.7114	921.1
220	20	38.385	1196.7	1717.8	16.794	5.3113	7.6688	936.35
230	20	36.842	1251.4	1794.3	17.134	5.3037	7.6316	951.39

(continued)

Table 5.6 (continued)

Temperature (K)	Pressure (MPa)	Density (kg/m^3)	Internal energy (kJ/kg)	Enthalpy (kJ/kg)	Entropy (kJ/kg K)	C_v (kJ/kg K)	C_p (kJ/kg K)	Sound speed (m/s)
240	20	35.424	1305.8	1870.4	17.458	5.2971	7.5989	966.22
250	20	34.117	1360.1	1946.3	17.768	5.2913	7.5701	980.84
260	20	32.908	1414.1	2021.8	18.064	5.2862	7.5445	995.26
270	20	31.785	1467.9	2097.2	18.348	5.2817	7.5217	1009.5
280	20	30.74	1521.7	2172.3	18.622	5.2777	7.5014	1023.5
290	20	29.765	1575.3	2247.2	18.885	5.2742	7.4833	1037.3
300	20	28.852	1628.7	2322	19.138	5.2711	7.4671	1051
400	20	22.134	2159.6	3063.2	21.271	5.2601	7.3757	1178.3
500	20	17.993	2688	3799.5	22.914	5.2784	7.3595	1292.1
600	20	15.171	3218.5	4536.8	24.258	5.3313	7.3955	1395.2
100.0 MPa								
40	50	187.07	40.984	308.27	0.78	4.6391	6.6185	1486.1
50	50	176.72	95.817	378.76	2.3494	5.2456	7.4596	1385.6
60	50	166.83	157.24	456.94	3.773	5.7349	8.144	1306.2
70	50	157.44	223.29	540.87	5.0657	6.0391	8.6006	1246
80	50	148.6	291.73	628.2	6.2315	6.1636	8.8315	1201.7
90	50	140.36	360.7	716.92	7.2765	6.1611	8.8914	1170
100	50	132.73	428.97	805.67	8.2116	6.0894	8.8451	1147.8
110	50	125.72	495.95	893.65	9.0502	5.9912	8.745	1133
120	50	119.31	561.42	980.51	9.8061	5.8917	8.6256	1123.5
130	50	113.44	625.41	1066.2	10.492	5.8029	8.5062	1118.2
140	50	108.09	688.07	1150.7	11.118	5.7287	8.3959	1116
150	50	103.19	749.58	1234.1	11.694	5.6685	8.2976	1116.3
160	50	98.708	810.11	1316.7	12.227	5.6204	8.2113	1118.6
170	50	94.594	869.81	1398.4	12.722	5.5819	8.1357	1122.5
180	50	90.809	928.79	1479.4	13.185	5.5508	8.0691	1127.6
190	50	87.319	987.18	1559.8	13.62	5.5251	8.01	1133.8
200	50	84.093	1045	1639.6	14.029	5.5035	7.957	1140.8
210	50	81.102	1102.4	1718.9	14.416	5.4851	7.9092	1148.5
220	50	78.323	1159.4	1797.8	14.783	5.469	7.8656	1156.8
230	50	75.735	1216.1	1876.3	15.132	5.4547	7.8258	1165.4
240	50	73.319	1272.4	1954.3	15.464	5.442	7.7894	1174.4
250	50	71.058	1328.4	2032.1	15.782	5.4306	7.7558	1183.7
260	50	68.939	1384.2	2109.5	16.085	5.4202	7.7249	1193.2
270	50	66.947	1439.7	2186.6	16.376	5.4109	7.6964	1202.9
280	50	65.073	1495	2263.4	16.656	5.4023	7.6701	1212.7
290	50	63.305	1550.2	2340	16.924	5.3946	7.6459	1222.6
300	50	61.635	1605.1	2416.3	17.183	5.3876	7.6235	1232.5
400	50	48.89	2147.7	3170.4	19.353	5.3472	7.4777	1331.9
500	50	40.619	2684	3914.9	21.015	5.3474	7.4251	1427.2
600	50	34.785	3220.2	4657.6	22.369	5.3882	7.4381	1516.5

References

1. P.C. Souers, *Hydrogen Properties for Fusion Energy* (University of California Press, Los Angeles, 1986)
2. I.A. Richardson, J.W. Leachman, E.W. Lemmon, Fundamental Equation of State for Deuterium. J. Phys. Chem. Ref. Data **43**(1), 0131031 (2014)

Chapter 6
Neon

6.1 Formulation for Neon

The available formulation for neon is one of the oldest within this text [14]. The uncertainty statement in REFPROP is as follows:

The uncertainties of the equation of state are 0.1% in density, 2% in heat capacity, and 1% in the speed of sound, except in the critical region. The uncertainty in vapor pressure is 0.2% (Tables 6.1, 6.2, 6.3 and 6.4).

J.W. Leachman et al., *Thermodynamic Properties of Cryogenic Fluids*,
International Cryogenics Monograph Series, DOI 10.1007/978-3-319-57835-4_6

Table 6.1 Fixed Point Properties and Correlation Limits for Neon

	Temperature (K)	Pressure (kPa)	Density (mol/dm^3)
Critical point	44.4918	2678.6	23.882
Triple point	24.556	43.464	
Upper limit	700.0	700,000	90.56
Lower limit	24.556		

Table 6.2 Fluid Constants and Normal Boiling Point for Neon

Molar mass	20.179 g/mol
Ideal-gas constant	8.31434 J/(mol K)
Pitzer's acentric factor	−0.0387
Normal boiling point	27.104 K

Table 6.3 Fundamental equation of state coefficients and exponents for neon

i	N_i	t_i	d_i	p_i
1	3.532653449	0.5	1	
2	−4.513954384	0.75	1	
3	−0.1524027959	3.5	1	
4	2.188568609	0.5	2	
5	−7.44299997	0.75	2	
6	7.755627402	1	2	
7	−3.122553128	1.5	2	
8	1.014206899	2.5	2	
9	−0.05289214086	0.25	3	
10	0.1566849239	0.5	3	
11	−0.222852705	2.5	3	
12	−0.01410150942	1	4	
13	0.07036229719	3	4	
14	−0.05882048367	4	4	
15	0.01571172741	5	4	
16	0.001292202769	1	6	
17	0.0007902035603	5	6	
18	−0.0003794403616	6	6	
19	0.04652799333	4	1	3

(continued)

Table 6.3 (continued)

i	N_i	t_i	d_i	p_i
20	0.04524001818	1	2	2
21	−0.2383421991	5	2	2
22	0.00629359013	8	2	4
23	−0.001272313644	12	2	6
24	−-0.000000175235256	32	2	6
25	0.007188419232	10	4	2
26	−0.05403006914	6	8	2
27	0.075782221874	7	8	2
28	−0.03808588254	8	8	2
29	0.006034022431	9	8	2

Table 6.4 Ancillary Function Coefficients and Exponents for Neon

	α^0
i	a_i
0	2.5
1	−3.0597521577
2	3.2525230449

$m = 0$, $n = 0$

	Vapor pressure		Saturated liquid density		Saturated vapor density	
i	N_i	k_i	N_i	k_i	N_i	k_i
1	−5.5805	1	1.0601	0.33	−2.3338	0.444
2	0.068795	1.5	120.76	1.4	−3.6834	0.95
3	5.484	2.3	−385.53	1.7	−85.368	3.5
4	−8.376	2.8	816.55	2.2	227.69	4.1
5	3.4276	3.4	−899.07	2.6	−172.9	4.5
6			354.66	3		

6.2 Property Plots for Neon

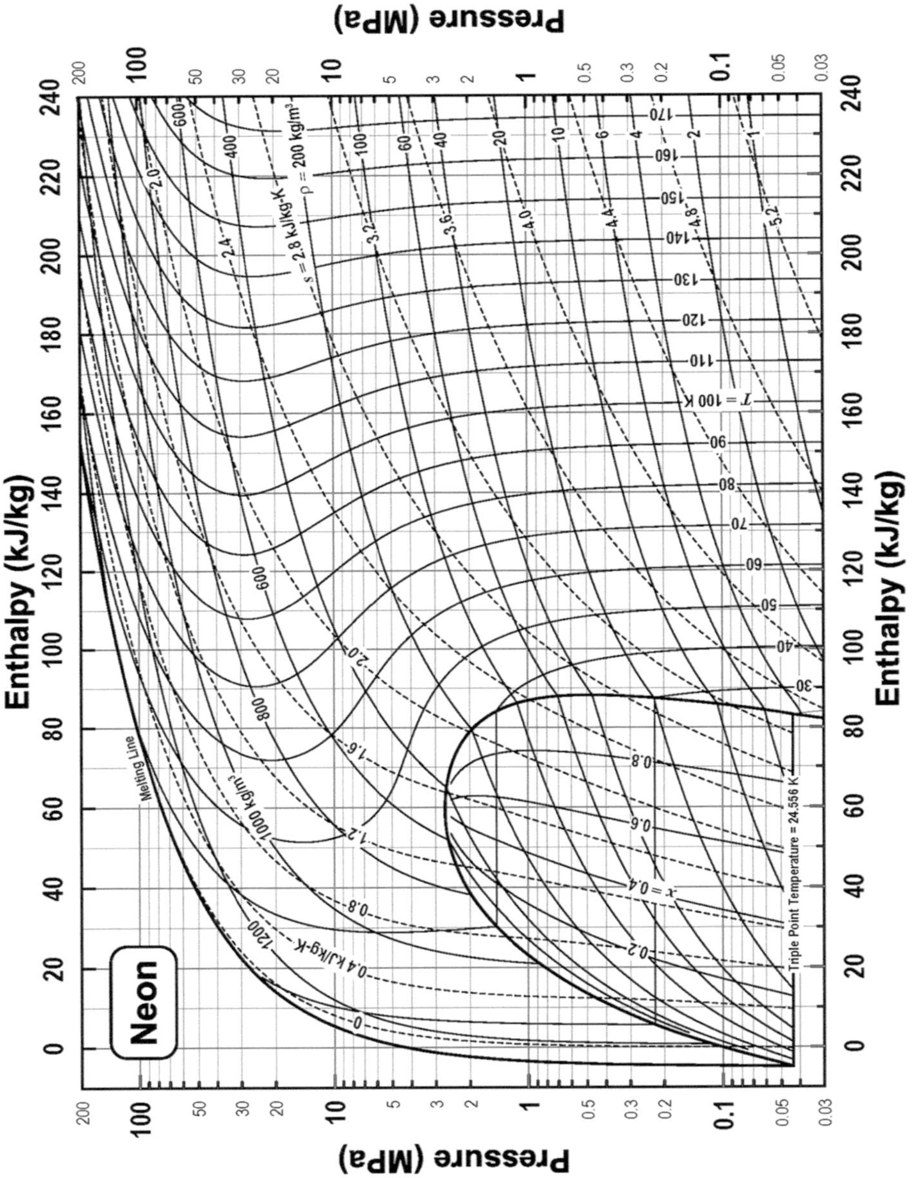
Neon
Enthalpy (kJ/kg)
Pressure (MPa)
Melting Line
Triple Point Temperature = 24.556 K
T = 100 K
s = 2.8 kJ/kg-K
ρ = 200 kg/m³
0.4 kJ/kg-K
1000 kg/m³
x = 0.4

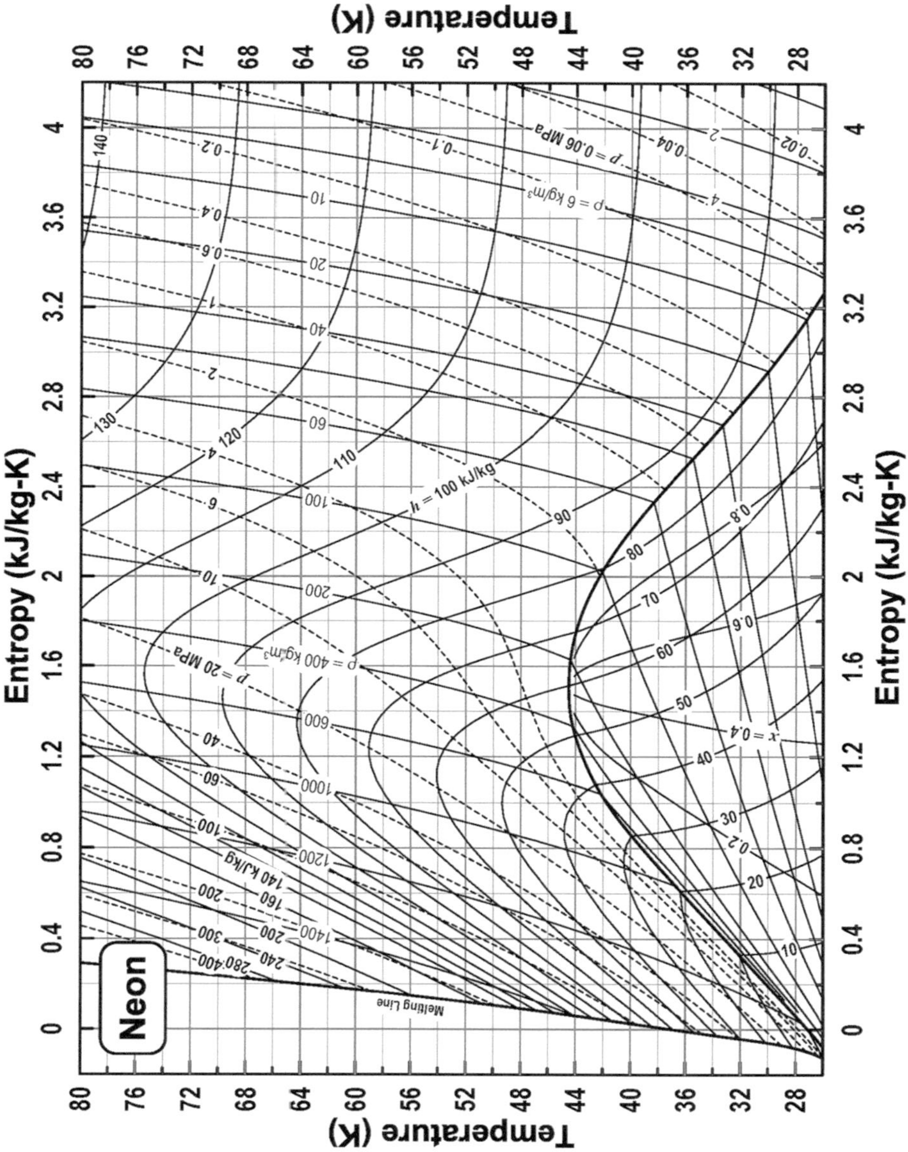
Neon
Entropy (kJ/kg-K)
Temperature (K)
Melting Line
h = 100 kJ/kg
p = 20 MPa
p = 0.06 MPa
ρ = 400 kg/m³
ρ = 6 kg/m³
x = 0.4

6.3 Thermodynamic Data Tables for Neon

See Tables 6.5 and 6.6.

Table 6.5 Thermodynamic Properties of Saturated Neon

Temperature (K)	Pressure (MPa)	Liquid density (kg/m³)	Vapor density (kg/m³)	Liquid enthalpy (kJ/kg)	Vapor enthalpy (kJ/kg)	Liquid entropy (kJ/kg K)	Vapor entropy (kJ/kg K)	Liquid C_v (kJ/kg K)	Vapor C_v (kJ/kg K)	Liquid sound speed (m/s)	Vapor sound speed (m/s)
24.556	0.043368	1252.4	4.4357	−4.9275	83.197	−0.18929	3.3994	0.92589	1.7479	675.01	112.98
25	0.05092	1243.9	5.1267	−4.0073	83.76	−0.15239	3.3583	0.91376	1.4777	652.08	115.48
26	0.071611	1226.1	6.9815	−2.0717	84.82	−0.07712	3.2649	0.90158	1.0924	617.51	121.12
27	0.098173	1208.8	9.3048	−0.19595	85.673	−0.00715	3.1732	0.89232	0.89354	596	126.17
27.104	0.10133	1207	9.5773	0	85.752	0	3.1638	0.89131	0.87949	594.15	126.64
28	0.13159	1191.2	12.168	1.6946	86.375	0.060592	3.0849	0.88225	0.79294	579.65	130.23
29	0.17287	1173	15.647	3.6332	86.957	0.12739	3.0006	0.87163	0.74439	564.56	133.28
30	0.22307	1154	19.828	5.6343	87.429	0.19376	2.9202	0.86084	0.72373	548.99	135.55
31	0.28324	1134	24.803	7.7035	87.793	0.25988	2.8434	0.85019	0.7184	532.26	137.25
32	0.35446	1113.1	30.68	9.8421	88.046	0.32576	2.7696	0.84007	0.72184	514.19	138.54
33	0.43782	1091.2	37.584	12.05	88.179	0.39138	2.6983	0.83096	0.73072	494.83	139.52
34	0.53443	1068.3	45.664	14.33	88.181	0.45677	2.6288	0.82332	0.74336	474.28	140.26
35	0.64543	1044.2	55.106	16.687	88.037	0.52203	2.5606	0.81757	0.759	452.65	140.78
36	0.77202	1018.9	66.143	19.13	87.727	0.58738	2.4929	0.81402	0.77743	429.96	141.11
37	0.91543	992.03	79.087	21.674	87.224	0.65318	2.4248	0.81295	0.79875	406.17	141.24
38	1.077	963.27	94.361	24.345	86.492	0.71999	2.3554	0.81461	0.82334	381.08	141.18
39	1.2581	932	112.58	27.179	85.479	0.78863	2.2835	0.81945	0.85184	354.35	140.93
40	1.4603	897.21	134.66	30.236	84.107	0.86043	2.2072	0.82833	0.88522	325.39	140.49
41	1.6855	857.03	162.17	33.623	82.251	0.9377	2.1238	0.84313	0.92493	293.27	139.83
42	1.9355	807.59	197.94	37.552	79.688	1.0251	2.0284	0.86831	0.973	256.54	138.95
43	2.2121	739.37	248.26	42.541	75.959	1.134	1.9112	0.91534	1.0319	213.78	137.87
44	2.5168	632.15	333.79	49.773	69.667	1.29	1.7421	1.0016	1.1005	170.57	136.83
44.492	2.6786	481.91	481.91	59.461	59.461	1.5022	1.5022				

Table 6.6 Thermodynamic Properties of Neon

Temperature (K)	Pressure (MPa)	Density (kg/m³)	Internal energy (kJ/kg)	Enthalpy (kJ/kg)	Entropy (kJ/kg K)	C_v (kJ/kg K)	C_p (kJ/kg K)	Sound speed (m/s)
0.1 MPa								
24.61	0.1	1251.6	−4.864	−4.7841	−0.1853	0.9263	2.1189	671.77
27.061	0.1	1207.7	−0.16463	−0.08183	−0.00298	0.89173	1.8613	594.92
27.061	0.1	9.4628	75.152	85.719	3.1677	0.88523	1.4208	126.45
30	0.1	8.4069	77.36	89.255	3.2919	0.64754	1.1202	140.59
40	0.1	6.1709	83.862	100.07	3.6034	0.62415	1.0614	164.59
50	0.1	4.8974	90.177	110.6	3.8384	0.62057	1.0468	184.77
60	0.1	4.0657	96.434	121.03	4.0286	0.61938	1.0406	202.77
70	0.1	3.4777	102.66	131.42	4.1888	0.61888	1.0374	219.21
80	0.1	3.0393	108.88	141.78	4.3272	0.61864	1.0354	234.46
90	0.1	2.6995	115.09	152.13	4.449	0.61851	1.0342	248.75
100	0.1	2.4283	121.29	162.47	4.5579	0.61843	1.0333	262.24
110	0.1	2.2069	127.48	172.8	4.6564	0.61837	1.0327	275.06
120	0.1	2.0225	133.68	183.12	4.7462	0.61833	1.0322	287.3
130	0.1	1.8666	139.87	193.44	4.8288	0.6183	1.0319	299.04
140	0.1	1.7331	146.06	203.76	4.9053	0.61828	1.0316	310.33
150	0.1	1.6174	152.25	214.07	4.9765	0.61826	1.0314	321.22
160	0.1	1.5162	158.44	224.39	5.043	0.61824	1.0312	331.75
170	0.1	1.427	164.62	234.7	5.1055	0.61823	1.0311	341.95
180	0.1	1.3477	170.81	245.01	5.1645	0.61822	1.031	351.85
190	0.1	1.2767	176.99	255.32	5.2202	0.61821	1.0309	361.49
200	0.1	1.2129	183.18	265.63	5.2731	0.6182	1.0308	370.87
210	0.1	1.1551	189.36	275.93	5.3234	0.61819	1.0307	380.02
220	0.1	1.1026	195.55	286.24	5.3713	0.61818	1.0306	388.95
230	0.1	1.0547	201.73	296.55	5.4171	0.61817	1.0306	397.69
240	0.1	1.0107	207.91	306.85	5.461	0.61817	1.0305	406.23
250	0.1	0.97031	214.1	317.16	5.5031	0.61816	1.0305	414.6
260	0.1	0.933	220.28	327.46	5.5435	0.61816	1.0305	422.8
270	0.1	0.89845	226.46	337.77	5.5824	0.61815	1.0304	430.85
280	0.1	0.86637	232.65	348.07	5.6198	0.61815	1.0304	438.74
290	0.1	0.8365	238.83	358.37	5.656	0.61814	1.0304	446.5
300	0.1	0.80862	245.01	368.68	5.6909	0.61814	1.0303	454.13
400	0.1	0.60652	306.83	471.71	5.9873	0.61811	1.0302	524.31
500	0.1	0.48525	368.64	574.72	6.2172	0.6181	1.0301	586.15
600	0.1	0.40439	430.45	677.74	6.405	0.61809	1.0301	642.06
700	0.1	0.34664	492.26	780.75	6.5638	0.61808	1.0301	693.47
0.2 MPa								
24.673	0.2	1250.9	−4.7631	−4.6032	−0.1812	0.9282	2.1003	668.17
29.564	0.2	1162.4	4.5807	4.7527	0.16483	0.86555	1.9831	555.91
29.564	0.2	17.911	76.07	87.236	2.9549	0.73031	1.3068	134.64
30	0.2	17.577	76.419	87.798	2.9737	0.70543	1.2686	136.63
40	0.2	12.562	83.357	99.278	3.305	0.63115	1.0967	163.4

(continued)

Table 6.6 (continued)

Temperature (K)	Pressure (MPa)	Density (kg/m^3)	Internal energy (kJ/kg)	Enthalpy (kJ/kg)	Entropy (kJ/kg K)	C_v (kJ/kg K)	C_p (kJ/kg K)	Sound speed (m/s)
50	0.2	9.8839	89.822	110.06	3.5456	0.62323	1.0645	184.23
60	0.2	8.1731	96.158	120.63	3.7384	0.62074	1.0515	202.55
70	0.2	6.9764	102.44	131.11	3.9	0.61973	1.0448	219.18
80	0.2	6.0895	108.69	141.53	4.0392	0.61924	1.0408	234.53
90	0.2	5.4046	114.92	151.93	4.1616	0.61897	1.0383	248.89
100	0.2	4.8593	121.14	162.3	4.2709	0.61881	1.0366	262.43
110	0.2	4.4146	127.36	172.66	4.3697	0.6187	1.0353	275.28
120	0.2	4.0449	133.56	183.01	4.4597	0.61862	1.0344	287.55
130	0.2	3.7326	139.77	193.35	4.5425	0.61856	1.0337	299.3
140	0.2	3.4652	145.97	203.68	4.6191	0.61851	1.0332	310.59
150	0.2	3.2337	152.16	214.01	4.6903	0.61847	1.0327	321.49
160	0.2	3.0312	158.36	224.34	4.757	0.61844	1.0324	332.02
170	0.2	2.8527	164.55	234.66	4.8195	0.61841	1.0321	342.22
180	0.2	2.6941	170.74	244.98	4.8785	0.61839	1.0319	352.13
190	0.2	2.5522	176.93	255.3	4.9343	0.61837	1.0317	361.76
200	0.2	2.4245	183.12	265.61	4.9872	0.61835	1.0315	371.14
210	0.2	2.309	189.31	275.93	5.0375	0.61833	1.0313	380.29
220	0.2	2.2041	195.5	286.24	5.0855	0.61832	1.0312	389.22
230	0.2	2.1083	201.69	296.55	5.1314	0.61831	1.0311	397.95
240	0.2	2.0204	207.87	306.86	5.1752	0.61829	1.031	406.49
250	0.2	1.9396	214.06	317.17	5.2173	0.61828	1.0309	414.85
260	0.2	1.8651	220.25	327.48	5.2578	0.61827	1.0308	423.05
270	0.2	1.796	226.43	337.79	5.2967	0.61826	1.0308	431.1
280	0.2	1.7319	232.62	348.1	5.3341	0.61825	1.0307	438.99
290	0.2	1.6722	238.8	358.4	5.3703	0.61824	1.0307	446.75
300	0.2	1.6165	244.98	368.71	5.4053	0.61824	1.0306	454.37
400	0.2	1.2126	306.82	471.76	5.7017	0.61818	1.0303	524.52
500	0.2	0.97018	368.64	574.78	5.9316	0.61815	1.0302	586.34
600	0.2	0.80857	430.45	677.8	6.1194	0.61813	1.0302	642.22
700	0.2	0.69311	492.26	780.82	6.2782	0.61811	1.0301	693.62
0.5 MPa								
24.817	0.5	1249.7	−4.5542	−4.1541	−0.17274	0.93457	2.0636	660.59
30	0.5	1156.1	5.3184	5.7509	0.18965	0.86247	2.0009	551.29
33.66	0.5	1076.2	13.082	13.546	0.43454	0.82572	2.2695	481.4
33.66	0.5	42.773	76.506	88.196	2.6523	0.7387	1.5245	140.04
40	0.5	33.306	81.703	96.716	2.8849	0.65865	1.2352	159.52
50	0.5	25.417	88.718	108.39	3.1458	0.63204	1.1235	182.63
60	0.5	20.753	95.317	119.41	3.3469	0.62493	1.0856	201.95
70	0.5	17.6	101.76	130.17	3.5127	0.62225	1.0675	219.11
80	0.5	15.305	108.12	140.78	3.6545	0.62101	1.0572	234.79
90	0.5	13.553	114.43	151.32	3.7786	0.62034	1.0507	249.35
100	0.5	12.168	120.71	161.81	3.8891	0.61993	1.0464	263.02

(continued)

Table 6.6 (continued)

Temperature (K)	Pressure (MPa)	Density (kg/m^3)	Internal energy (kJ/kg)	Enthalpy (kJ/kg)	Entropy (kJ/kg K)	C_v (kJ/kg K)	C_p (kJ/kg K)	Sound speed (m/s)
110	0.5	11.043	126.98	172.25	3.9887	0.61966	1.0432	275.96
120	0.5	10.111	133.22	182.67	4.0793	0.61947	1.0409	288.28
130	0.5	9.3263	139.46	193.07	4.1626	0.61932	1.0392	300.07
140	0.5	8.6553	145.69	203.46	4.2396	0.61921	1.0378	311.39
150	0.5	8.0752	151.91	213.83	4.3111	0.61911	1.0367	322.3
160	0.5	7.5684	158.13	224.19	4.378	0.61903	1.0358	332.84
170	0.5	7.1218	164.34	234.55	4.4408	0.61896	1.0351	343.05
180	0.5	6.7253	170.55	244.9	4.4999	0.61891	1.0345	352.95
190	0.5	6.3708	176.76	255.24	4.5558	0.61885	1.034	362.58
200	0.5	6.052	182.96	265.58	4.6089	0.61881	1.0336	371.95
210	0.5	5.7636	189.16	275.91	4.6593	0.61877	1.0332	381.1
220	0.5	5.5016	195.36	286.24	4.7073	0.61873	1.0329	390.02
230	0.5	5.2624	201.56	296.57	4.7532	0.6187	1.0326	398.74
240	0.5	5.0433	207.75	306.89	4.7972	0.61866	1.0324	407.27
250	0.5	4.8417	213.95	317.22	4.8393	0.61864	1.0322	415.63
260	0.5	4.6556	220.14	327.54	4.8798	0.61861	1.032	423.81
270	0.5	4.4833	226.33	337.86	4.9188	0.61859	1.0318	431.85
280	0.5	4.3234	232.52	348.17	4.9563	0.61856	1.0317	439.73
290	0.5	4.1745	238.71	358.49	4.9925	0.61854	1.0316	447.48
300	0.5	4.0355	244.9	368.81	5.0274	0.61852	1.0314	455.09
400	0.5	3.0279	306.78	471.91	5.3241	0.61839	1.0307	525.15
500	0.5	2.4231	368.62	574.97	5.554	0.6183	1.0304	586.89
600	0.5	2.0198	430.45	678	5.7419	0.61825	1.0303	642.72
700	0.5	1.7316	492.27	781.02	5.9007	0.61822	1.0302	694.07
1.0 MPa								
25.011	1	1248.6	−4.3072	−3.5063	−0.16281	0.94413	2.0233	652.25
30	1	1159.9	5.1025	5.9647	0.18239	0.8651	1.9823	555.49
37.538	1	976.83	22.07	23.093	0.68896	0.81348	2.7146	392.85
37.538	1	86.982	75.365	86.862	2.3877	0.81154	2.1172	141.23
40	1	75.681	78.239	91.453	2.5063	0.73671	1.6807	151.55
50	1	53.498	86.736	105.43	2.8198	0.65012	1.2478	179.95
60	1	42.628	93.869	117.33	3.0371	0.63242	1.149	201.07
70	1	35.737	100.6	128.59	3.2107	0.6265	1.1075	219.13
80	1	30.88	107.16	139.54	3.357	0.62392	1.0853	235.31
90	1	27.24	113.61	150.32	3.484	0.62258	1.0717	250.19
100	1	24.396	120	160.99	3.5964	0.62177	1.0627	264.06
110	1	22.105	126.35	171.58	3.6974	0.62124	1.0564	277.13
120	1	20.218	132.66	182.12	3.7891	0.62086	1.0517	289.54
130	1	18.634	138.96	192.62	3.8731	0.62057	1.0482	301.39
140	1	17.284	145.23	203.09	3.9507	0.62035	1.0455	312.74
150	1	16.12	151.5	213.53	4.0228	0.62016	1.0433	323.67
160	1	15.104	157.75	223.96	4.09	0.62001	1.0415	334.22

(continued)

Table 6.6 (continued)

Temperature (K)	Pressure (MPa)	Density (kg/m^3)	Internal energy (kJ/kg)	Enthalpy (kJ/kg)	Entropy (kJ/kg K)	C_v (kJ/kg K)	C_p (kJ/kg K)	Sound speed (m/s)
170	1	14.21	163.99	234.36	4.1531	0.61987	1.0401	344.43
180	1	13.418	170.23	244.76	4.2125	0.61976	1.0389	354.33
190	1	12.709	176.46	255.14	4.2687	0.61965	1.0379	363.95
200	1	12.073	182.69	265.52	4.3219	0.61956	1.037	373.32
210	1	11.497	188.91	275.88	4.3725	0.61948	1.0363	382.45
220	1	10.975	195.12	286.24	4.4207	0.61941	1.0357	391.35
230	1	10.498	201.34	296.6	4.4667	0.61934	1.0352	400.06
240	1	10.061	207.55	306.95	4.5107	0.61928	1.0347	408.58
250	1	9.6586	213.76	317.29	4.553	0.61922	1.0343	416.91
260	1	9.2877	219.96	327.63	4.5935	0.61917	1.0339	425.09
270	1	8.9444	226.17	337.97	4.6325	0.61912	1.0336	433.1
280	1	8.6256	232.37	348.3	4.6701	0.61908	1.0333	440.97
290	1	8.3289	238.57	358.64	4.7064	0.61904	1.033	448.69
300	1	8.0519	244.77	368.96	4.7414	0.619	1.0328	456.29
400	1	6.044	306.71	472.16	5.0383	0.61873	1.0314	526.19
500	1	4.8385	368.6	575.27	5.2684	0.61856	1.0308	587.81
600	1	4.0342	430.45	678.33	5.4563	0.61846	1.0305	643.54
700	1	3.4592	492.29	781.37	5.6151	0.61839	1.0303	694.81
2.0 MPa								
25.336	2	1247.7	−3.9512	−2.3483	−0.14864	0.95748	1.9702	642.95
30	2	1167.1	4.6911	6.4047	0.16841	0.86926	1.9485	564.02
40	2	916.07	27.551	29.735	0.83301	0.82353	3.0965	346.17
42.242	2	793.37	36.111	38.632	1.0488	0.877	5.8987	246.74
42.242	2	208.42	69.322	78.918	2.0025	0.9862	6.0364	138.71
50	2	121.01	82.024	98.552	2.4361	0.70377	1.6668	174.69
60	2	90.146	90.782	112.97	2.6999	0.64959	1.3051	199.97
70	2	73.644	98.233	125.39	2.8916	0.63517	1.1955	219.7
80	2	62.802	105.22	137.06	3.0475	0.6296	1.144	236.77
90	2	54.977	111.96	148.34	3.1804	0.62688	1.1145	252.17
100	2	49.003	118.57	159.39	3.2968	0.62531	1.0955	266.37
110	2	44.265	125.09	170.27	3.4006	0.62429	1.0825	279.65
120	2	40.402	131.54	181.05	3.4943	0.62356	1.073	292.19
130	2	37.183	137.95	191.74	3.5799	0.62301	1.066	304.12
140	2	34.456	144.33	202.37	3.6587	0.62258	1.0605	315.53
150	2	32.112	150.67	212.95	3.7317	0.62222	1.0562	326.48
160	2	30.074	157	223.5	3.7998	0.62191	1.0527	337.04
170	2	28.286	163.3	234.01	3.8635	0.62165	1.0498	347.24
180	2	26.702	169.59	244.5	3.9235	0.62143	1.0475	357.13
190	2	25.289	175.87	254.96	3.98	0.62123	1.0455	366.73
200	2	24.02	182.14	265.41	4.0336	0.62105	1.0438	376.07
210	2	22.874	188.41	275.84	4.0845	0.62089	1.0424	385.17
220	2	21.834	194.66	286.26	4.133	0.62074	1.0412	394.04

(continued)

Table 6.6 (continued)

Temperature (K)	Pressure (MPa)	Density (kg/m^3)	Internal energy (kJ/kg)	Enthalpy (kJ/kg)	Entropy (kJ/kg K)	C_v (kJ/kg K)	C_p (kJ/kg K)	Sound speed (m/s)
230	2	20.886	200.91	296.66	4.1792	0.62061	1.0401	402.71
240	2	20.017	207.15	307.06	4.2235	0.62049	1.0392	411.19
250	2	19.219	213.38	317.45	4.2659	0.62038	1.0384	419.5
260	2	18.482	219.61	327.83	4.3066	0.62028	1.0376	427.63
270	2	17.8	225.84	338.2	4.3458	0.62019	1.037	435.61
280	2	17.167	232.07	348.57	4.3835	0.6201	1.0364	443.44
290	2	16.578	238.29	358.93	4.4198	0.62002	1.0359	451.14
300	2	16.028	244.51	369.29	4.4549	0.61994	1.0355	458.69
400	2	12.041	306.58	472.68	4.7524	0.6194	1.0327	528.28
500	2	9.6462	368.54	575.88	4.9827	0.61908	1.0315	589.65
600	2	8.0467	430.45	679	5.1707	0.61887	1.0309	645.18
700	2	6.9026	492.32	782.07	5.3296	0.61873	1.0306	696.29
5.0 MPa								
26.124	5	1249.1	−3.2642	0.73871	−0.12205	0.96554	1.876	639.94
30	5	1186.8	3.5911	7.8043	0.13011	0.87569	1.8671	590.18
40	5	982.38	23.762	28.852	0.73218	0.81807	2.4005	420.22
50	5	554.41	55.456	64.475	1.5141	0.88538	5.2941	201.36
60	5	268.96	79.716	98.306	2.1409	0.71595	2.09	206.03
70	5	199.54	90.661	115.72	2.4104	0.66239	1.5206	227.22
80	5	163.51	99.284	129.86	2.5995	0.64572	1.3345	245.03
90	5	140.15	107.04	142.71	2.751	0.63872	1.2446	260.84
100	5	123.33	114.35	154.88	2.8792	0.63502	1.1921	275.27
110	5	110.53	121.4	166.62	2.9911	0.63271	1.1579	288.68
120	5	100.41	128.27	178.07	3.0907	0.63109	1.134	301.27
130	5	92.09	135.02	189.32	3.1808	0.62985	1.1164	313.19
140	5	85.133	141.68	200.41	3.263	0.62886	1.1031	324.56
150	5	79.214	148.27	211.39	3.3387	0.62804	1.0926	335.45
160	5	74.107	154.8	222.27	3.409	0.62735	1.0842	345.92
170	5	69.649	161.29	233.08	3.4745	0.62675	1.0774	356.02
180	5	65.719	167.74	243.82	3.5359	0.62622	1.0718	365.8
190	5	62.226	174.16	254.52	3.5937	0.62575	1.0671	375.29
200	5	59.098	180.56	265.17	3.6484	0.62533	1.0631	384.5
210	5	56.279	186.94	275.78	3.7002	0.62495	1.0597	393.48
220	5	53.725	193.3	286.36	3.7494	0.62461	1.0568	402.24
230	5	51.398	199.64	296.92	3.7963	0.6243	1.0543	410.78
240	5	49.269	205.97	307.45	3.8411	0.62401	1.0521	419.14
250	5	47.314	212.28	317.96	3.884	0.62375	1.0501	427.32
260	5	45.511	218.59	328.45	3.9252	0.6235	1.0484	435.34
270	5	43.843	224.89	338.93	3.9647	0.62328	1.0469	443.2
280	5	42.295	231.17	349.39	4.0028	0.62307	1.0455	450.92
290	5	40.854	237.45	359.84	4.0394	0.62287	1.0443	458.5
300	5	39.51	243.73	370.28	4.0748	0.62269	1.0432	465.95

(continued)

Table 6.6 (continued)

Temperature (K)	Pressure (MPa)	Density (kg/m³)	Internal energy (kJ/kg)	Enthalpy (kJ/kg)	Entropy (kJ/kg K)	C_v (kJ/kg K)	C_p (kJ/kg K)	Sound speed (m/s)
400	5	29.758	306.2	474.22	4.3739	0.62138	1.0366	534.57
500	5	23.887	368.39	577.71	4.6048	0.6206	1.0336	595.18
600	5	19.956	430.45	680.99	4.7931	0.62009	1.0321	650.12
700	5	17.139	492.42	784.16	4.9522	0.61974	1.0313	700.76
10.0 MPa								
27.205	10	1255.3	−2.5553	5.4112	−0.09656	0.93821	1.7815	663.34
30	10	1214.5	2.0863	10.32	0.075199	0.8768	1.775	633.66
40	10	1046.5	20.064	29.62	0.62843	0.82522	2.0678	494.41
50	10	832.13	40.378	52.396	1.1347	0.78211	2.5388	360.17
60	10	585.21	61.783	78.87	1.6172	0.74404	2.5797	277.36
70	10	421.49	78.044	101.77	1.9714	0.69427	1.9949	264.98
80	10	333.98	89.692	119.63	2.2105	0.66769	1.6225	272.94
90	10	280.82	99.222	134.83	2.3897	0.65541	1.4366	284.48
100	10	244.42	107.72	148.63	2.5352	0.64895	1.332	296.54
110	10	217.52	115.63	161.6	2.6589	0.64496	1.266	308.41
120	10	196.63	123.16	174.02	2.767	0.64216	1.221	319.92
130	10	179.82	130.45	186.06	2.8634	0.64003	1.1885	331.03
140	10	165.92	137.55	197.82	2.9505	0.6383	1.164	341.73
150	10	154.21	144.51	209.36	3.0301	0.63686	1.1449	352.07
160	10	144.17	151.36	220.73	3.1035	0.63561	1.1298	362.06
170	10	135.45	158.13	231.96	3.1716	0.63453	1.1174	371.75
180	10	127.79	164.83	243.08	3.2352	0.63357	1.1072	381.14
190	10	121.01	171.48	254.11	3.2949	0.63272	1.0987	390.28
200	10	114.95	178.07	265.06	3.351	0.63195	1.0915	399.17
210	10	109.5	184.63	275.95	3.4041	0.63125	1.0853	407.84
220	10	104.57	191.14	286.77	3.4545	0.63062	1.08	416.31
230	10	100.08	197.63	297.55	3.5024	0.63004	1.0754	424.59
240	10	95.983	204.1	308.28	3.5481	0.6295	1.0714	432.69
250	10	92.217	210.54	318.98	3.5917	0.62901	1.0678	440.62
260	10	88.745	216.96	329.64	3.6336	0.62856	1.0647	448.4
270	10	85.534	223.36	340.27	3.6737	0.62814	1.0619	456.04
280	10	82.555	229.75	350.88	3.7122	0.62775	1.0594	463.54
290	10	79.782	236.12	361.46	3.7494	0.62738	1.0571	470.91
300	10	77.194	242.48	372.02	3.7852	0.62704	1.0551	478.16
400	10	58.395	305.57	476.82	4.0868	0.62455	1.0427	545.08
500	10	47.029	368.14	580.77	4.3187	0.62306	1.0371	604.43
600	10	39.388	430.44	684.32	4.5075	0.62208	1.0342	658.38
700	10	33.891	492.59	787.65	4.6668	0.62138	1.0325	708.23
50.0 MPa								
33.465	50	1310.2	0.41475	38.578	−0.0295	0.93171	1.5946	839.7
40	50	1250.6	9.1704	49.151	0.25899	0.91864	1.6314	769.46
50	50	1159.5	22.303	65.424	0.62216	0.84822	1.6212	698.34

(continued)

Table 6.6 (continued)

Temperature (K)	Pressure (MPa)	Density (kg/m³)	Internal energy (kJ/kg)	Enthalpy (kJ/kg)	Entropy (kJ/kg K)	C_v (kJ/kg K)	C_p (kJ/kg K)	Sound speed (m/s)
60	50	1070.9	34.92	81.609	0.91726	0.80509	1.6161	644.56
70	50	987.24	47.059	97.705	1.1654	0.7792	1.5997	602.1
80	50	910.63	58.626	113.53	1.3768	0.75975	1.5632	570.1
90	50	842.18	69.562	128.93	1.5583	0.74389	1.5156	547.25
100	50	781.81	79.889	143.84	1.7154	0.73077	1.4671	531.65
110	50	728.8	89.681	158.29	1.8531	0.71985	1.4225	521.46
120	50	682.19	99.016	172.31	1.9751	0.71071	1.3831	515.22
130	50	641.07	107.97	185.96	2.0845	0.703	1.3486	511.89
140	50	604.63	116.6	199.3	2.1833	0.69646	1.3186	510.73
150	50	572.18	124.96	212.35	2.2733	0.69087	1.2922	511.19
160	50	543.15	133.1	225.15	2.356	0.68607	1.269	512.87
170	50	517.05	141.03	237.74	2.4323	0.6819	1.2486	515.47
180	50	493.46	148.81	250.13	2.5031	0.67825	1.2305	518.78
190	50	472.06	156.44	262.35	2.5692	0.67504	1.2144	522.6
200	50	452.55	163.94	274.42	2.6311	0.67218	1.2001	526.83
210	50	434.69	171.34	286.36	2.6894	0.66962	1.1872	531.36
220	50	418.29	178.64	298.17	2.7443	0.66731	1.1757	536.11
230	50	403.15	185.86	309.88	2.7964	0.66521	1.1654	541.03
240	50	389.15	193	321.48	2.8458	0.6633	1.156	546.08
250	50	376.16	200.08	333	2.8928	0.66154	1.1475	551.22
260	50	364.06	207.1	344.44	2.9376	0.65992	1.1398	556.41
270	50	352.76	214.06	355.8	2.9805	0.65842	1.1327	561.66
280	50	342.2	220.98	367.09	3.0216	0.65703	1.1263	566.93
290	50	332.28	227.85	378.33	3.061	0.65573	1.1203	572.21
300	50	322.96	234.68	389.5	3.0989	0.65452	1.1149	577.5
400	50	253.14	301.42	498.94	3.414	0.64554	1.0782	629.46
500	50	208.9	366.37	605.72	3.6523	0.63994	1.0592	678.61
600	50	178.12	430.33	711.05	3.8444	0.63608	1.0483	724.87
700	50	155.37	493.7	815.51	4.0054	0.63326	1.0415	768.6
100.0 MPa								
39.563	100	1363.5	4.5136	77.852	0.022402	1.0225	1.5271	947.89
40	100	1360.8	5.0332	78.52	0.039183	1.0188	1.5263	944.69
50	100	1298	16.619	93.66	0.37714	0.93421	1.5019	887.27
60	100	1236.3	27.733	108.62	0.64988	0.88499	1.4923	842.4
70	100	1177.1	38.566	123.52	0.87964	0.85912	1.4879	804.21
80	100	1121.1	49.148	138.35	1.0776	0.83995	1.4756	772.82
90	100	1068.8	59.439	153	1.2503	0.8222	1.4544	747.78
100	100	1020.3	69.409	167.42	1.4022	0.80526	1.4277	728.18
110	100	975.63	79.054	181.55	1.5369	0.7895	1.3991	713.02
120	100	934.47	88.39	195.4	1.6574	0.77526	1.3712	701.39
130	100	896.56	97.444	208.98	1.7661	0.76263	1.3451	692.56
140	100	861.59	106.25	222.31	1.8649	0.75157	1.3214	685.94

(continued)

Table 6.6 (continued)

Temperature (K)	Pressure (MPa)	Density (kg/m^3)	Internal energy (kJ/kg)	Enthalpy (kJ/kg)	Entropy (kJ/kg K)	C_v (kJ/kg K)	C_p (kJ/kg K)	Sound speed (m/s)
150	100	829.26	114.83	235.42	1.9554	0.74191	1.3001	681.09
160	100	799.31	123.21	248.32	2.0387	0.7335	1.2812	677.65
170	100	771.5	131.43	261.05	2.1158	0.72615	1.2644	675.35
180	100	745.6	139.5	273.62	2.1877	0.71971	1.2494	674
190	100	721.43	147.43	286.04	2.2548	0.71404	1.2359	673.43
200	100	698.83	155.24	298.34	2.3179	0.70903	1.2238	673.52
210	100	677.65	162.95	310.52	2.3774	0.70457	1.2127	674.15
220	100	657.76	170.57	322.6	2.4335	0.70059	1.2026	675.25
230	100	639.05	178.1	334.58	2.4868	0.697	1.1934	676.75
240	100	621.43	185.55	346.47	2.5374	0.69376	1.1849	678.59
250	100	604.79	192.93	358.28	2.5856	0.69081	1.177	680.71
260	100	589.05	200.25	370.01	2.6316	0.68812	1.1697	683.1
270	100	574.16	207.5	381.67	2.6756	0.68565	1.1629	685.69
280	100	560.04	214.71	393.27	2.7178	0.68337	1.1565	688.47
290	100	546.63	221.86	404.8	2.7583	0.68126	1.1505	691.42
300	100	533.89	228.97	416.28	2.7972	0.67931	1.1449	694.5
400	100	434.09	298.13	528.5	3.1203	0.66514	1.1032	729.7
500	100	366.96	364.96	637.47	3.3636	0.65636	1.0782	767.75
600	100	318.45	430.41	744.43	3.5586	0.65022	1.0621	805.72
700	100	281.6	494.96	850.07	3.7215	0.64563	1.0514	842.83

Reference

1. R.S. Katti, R.T. Jacobsen, R.B. Stewart, M. Jahangiri, Thermodynamic properties for neon for temperatures from the triple point to 700 K at pressures to 700 MPa. Adv. Cryog. Eng. **31**, 1189–1197 (1986)

Chapter 7
Nitrogen

7.1 Formulation for Nitrogen

Due to its abundance and role in industry as a cryogenic fluid, nitrogen is one of the most well measured fluids at states in the vapor and liquid at cryogenic conditions, the region very near the critical point, and at temperatures and pressures far in excess of those typically required in most applications. The critical region has been extensively measured by multiple sources and with extremely low uncertainties, and this has led to the development of a high-accuracy equation of state by Span et al., published in 2000, that has been used in a number of critical region studies. Nitrogen is a reference fluid for many applications such as flow meter calibration, and thus the requirements on accuracy of the equation are high, as detailed below. The formulation for nitrogen was a collaboration between Span, Lemmon, Jacobsen, Wagner, and Yokozeki during 1996, during which Roland Span spent six months in Idaho working with the researchers at CATS [1]. That collaboration continues today, and has resulted in the development of more than 25 new equations of state between CATS, NIST, and the Ruhr University in Bochum. The uncertainties in properties in the equation of state are as follows.

The uncertainty in density of the equation of state is 0.02% from the triple point up to temperatures of 523 K and pressures up to 12 MPa and from temperatures of 240 to 523 K at pressures less than 30 MPa. In the range from 270 to 350 K at pressures less than 12 MPa, the uncertainty in density is 0.01%. The uncertainty at very high pressures (>1 GPa) is 0.6% in density. The uncertainty in pressure in the critical region is estimated to be 0.02%. In the gaseous and supercritical region, the speed of sound can be calculated with a typical uncertainty of 0.005–0.1%. At liquid states and at high pressures, the uncertainty increases to 0.5–1.5%. For pressures up to 30 MPa, the estimated uncertainty for heat capacities ranges from 0.3% at gaseous and gas like supercritical states up to 0.8% at liquid states and at certain gaseous and supercritical states at low temperatures. The uncertainty is 2% for pressures up to 200 MPa and larger at higher pressures. The estimated uncertainties of vapor pressure, saturated liquid density, and saturated vapor density are in general 0.02% for each property.

J.W. Leachman et al., *Thermodynamic Properties of Cryogenic Fluids*,
International Cryogenics Monograph Series, DOI 10.1007/978-3-319-57835-4_7

The formulation yields a reasonable extrapolation behavior up to the limits of chemical stability of nitrogen (Tables 7.1, 7.2, 7.3 and 7.4).

The functional form of the ancillary equations for nitrogen are different than those of the other fluids, see Chap. 2 for the correct form.

Table 7.1 Fixed point properties and correlation limits for nitrogen

	Temperature (K)	Pressure (kPa)	Density (mol/dm^3)
Critical point	126.192	3395.8	11.1839
Triple point	63.151	12.5198	
Upper limit	2000	2,200,000	53.15
Lower limit	63.151		

Table 7.2 Fluid constants and normal boiling point for nitrogen

Molar mass	28.01348 g/mol
Ideal-gas constant	8.31451 J/(mol K)
Pitzer's acentric factor	0.0372
Normal boiling point	77.355 K

Table 7.3 Fundamental equation of state coefficients and exponents for nitrogen

i	N_i	t_i	d_i	p_i	φ_i	β_i	γ_i	D_i
1	0.924803575	0.25	1					
2	−0.492448489	0.875	1					
3	0.661883337	0.5	2					
4	−1.929026492	0.875	2					
5	−0.062246931	0.375	3					
6	0.349943958	0.75	3					
7	0.564857472	0.5	1	1				
8	−1.61720006	0.75	1	1				
9	−0.481395032	2	1	1				
10	0.421150636	1.25	3	1				
11	−0.016196223	3.5	3	1				
12	0.172100994	1	4	1				
13	0.007354489	0.5	6	1				
14	0.016807731	3	6	1				
15	−0.001076267	0	7	1				
16	−0.013731809	2.75	7	1				
17	0.000635467	0.75	8	1				
18	0.003044323	2.5	8	1				
19	−0.043576234	4	1	2				
20	−0.072317489	6	2	2				
21	0.038964432	6	3	2				

(continued)

Table 7.3 (continued)

i	N_i	t_i	d_i	p_i	φ_i	β_i	γ_i	D_i
22	−0.021220136	3	4	2				
23	0.00408823	3	5	2				
24	−0.000055199	6	8	2				
25	−0.046201672	16	4	3				
26	−0.003003117	11	5	3				
27	0.036882589	15	5	3				
28	−0.002558568	12	8	3				
29	0.008969153	12	3	4				
30	−0.004415134	7	5	4				
31	0.001337229	4	6	4				
32	0.000264832	16	9	4				
33	19.6688194	0	1		20	325	1.16	1
34	−20.91156007	1	1		20	325	1.16	1
35	0.016778831	2	3		15	300	1.13	1
36	2627.675663	3	2		25	275	1.25	1

Table 7.4 Ancillary function coefficients and exponents for nitrogen

	α^0	
i	a_i	k_i
0	3.5	
1	−12.7695270805	
2	−0.0078416297	
3	3.0664690×10^{-6}	1
4	4.701240×10^{-9}	2
5	-3.9879840×10^{-13}	3
6	1.012941	3364.011

$m = 5$, $n = 6$

	Vapor pressure		Saturated liquid density		Saturated vapor density	
i	N_i	k_i	N_i	k_i	N_i	k_i
1	−6.12445	1	1.48654237	0.3294	−1.70127	0.34
2	1.263272	1.5	−0.280476066	2/3	−3.70403	2.5/3
3	−0.76591	2.5	0.089414309	8/3	1.298594	3.5/3
4	−1.77571	5	−0.119879866	17.5/3	−0.56142	6.5/3
5					−2.68505	14/3

7.2 Property Plots for Nitrogen

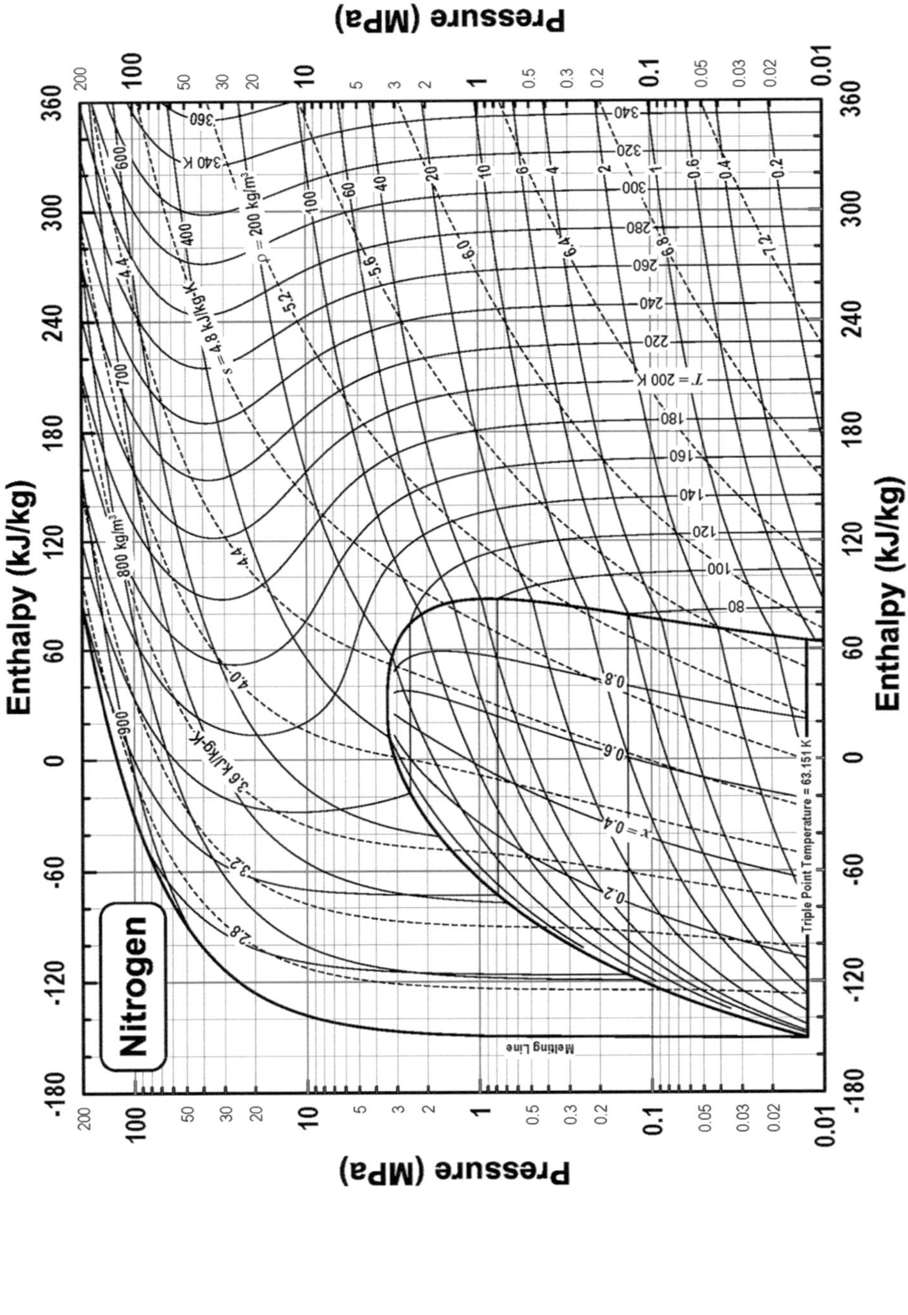
Nitrogen
Enthalpy (kJ/kg)
Pressure (MPa)
Melting Line
Triple Point Temperature = 63.151 K
s = 4.8 kJ/kg-K
3.6 kJ/kg-K
ρ = 200 kg/m³
800 kg/m³
T = 200 K
340 K
x = 0.4

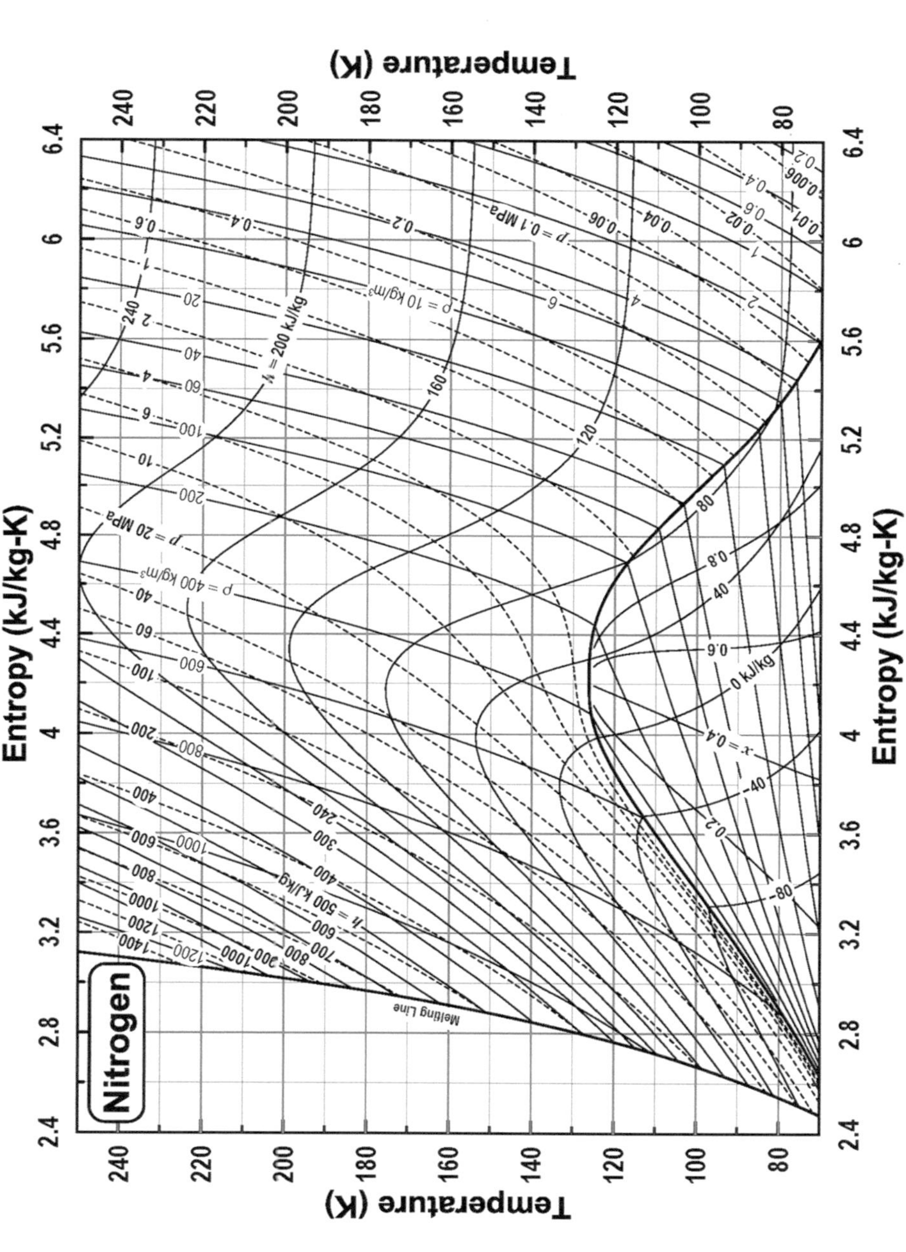
Nitrogen
Temperature (K)
Entropy (kJ/kg-K)
Melting Line
p = 20 MPa
p = 0.1 MPa
ρ = 400 kg/m³
ρ = 10 kg/m³
h = 200 kJ/kg
h = 500 kJ/kg
0 kJ/kg
x = 0.4

7.3 Thermodynamic Data Tables for Nitrogen

See Tables 7.5 and 7.6.

Table 7.5 Thermodynamic properties of saturated nitrogen

Temperature (K)	Pressure (MPa)	Liquid density (kg/m³)	Vapor density (kg/m³)	Liquid enthalpy (kJ/kg)	Vapor enthalpy (kJ/kg)	Liquid entropy (kJ/kg K)	Vapor entropy (kJ/kg K)	Liquid C_v (kJ/kg K)	Vapor C_v (kJ/kg K)	Liquid sound speed (m/s)	Vapor sound speed (m/s)
63.151	0.01252	867.22	0.67427	−150.73	64.779	2.4257	5.8383	1.1763	0.7499	995.28	161.11
66	0.020623	855.44	1.0673	−145.02	67.472	2.514	5.7336	1.1565	0.75284	966.18	164.3
69	0.033213	842.79	1.6537	−138.99	70.21	2.6032	5.635	1.1363	0.75663	935.83	167.43
72	0.051213	829.88	2.4616	−132.93	72.832	2.6889	5.5466	1.1167	0.76116	905.58	170.32
75	0.076043	816.67	3.5404	−126.83	75.316	2.7714	5.4667	1.0981	0.76649	875.28	172.95
77.355	0.10133	806.08	4.6121	−122.02	77.158	2.8342	5.409	1.0841	0.77128	851.39	174.82
78	0.10926	803.15	4.9442	−120.7	77.644	2.8511	5.3939	1.0803	0.77268	844.82	175.31
81	0.15251	789.27	6.7317	−114.51	79.795	2.9282	5.327	1.0636	0.77981	814.07	177.38
84	0.20757	774.99	8.9672	−108.26	81.749	3.0031	5.2651	1.0479	0.78797	782.93	179.16
87	0.27626	760.26	11.723	−101.93	83.482	3.0761	5.2073	1.0332	0.79727	751.28	180.63
90	0.36046	745.02	15.079	−95.517	84.97	3.1473	5.1527	1.0196	0.80785	719.01	181.78
93	0.4621	729.19	19.131	−88.992	86.185	3.2171	5.1007	1.0072	0.81986	685.99	182.62
96	0.58316	712.67	23.989	−82.336	87.095	3.2858	5.0507	0.99596	0.83353	652.09	183.12
99	0.72566	695.34	29.79	−75.522	87.662	3.3536	5.0019	0.9861	0.84912	617.14	183.28
102	0.89166	677.03	36.705	−68.514	87.837	3.4209	4.9538	0.97783	0.86701	580.96	183.08
105	1.0833	657.52	44.959	−61.268	87.557	3.4882	4.9055	0.97148	0.88744	543.3	182.51
108	1.3028	636.5	54.857	−53.723	86.739	3.5558	4.8564	0.96758	0.91066	503.88	181.58
111	1.5526	613.55	66.84	−45.793	85.258	3.6246	4.8052	0.96699	0.93825	462.32	180.28
114	1.8351	587.98	81.587	−37.35	82.925	3.6954	4.7505	0.97123	0.97454	418.12	178.49
117	2.1533	558.66	100.25	−28.181	79.431	3.77	4.6898	0.98327	1.0253	370.43	176.01
120	2.5106	523.36	125.09	−17.87	74.173	3.8514	4.6185	1.0106	1.0985	317.33	172.61
123	2.9116	476.14	162.05	−5.3412	65.597	3.9479	4.5246	1.0822	1.2203	253.32	167.43
126	3.3645	372.04	255.22	17.576	42.656	4.1232	4.3222	1.5491	1.6936	150.97	148.38
126.19	3.3955	323.04	303.47	27.209	31.348	4.1988	4.2316				

Table 7.6 Thermodynamic properties of nitrogen

Temperature (K)	Pressure (MPa)	Density (kg/m³)	Internal energy (kJ/kg)	Enthalpy (kJ/kg)	Entropy (kJ/kg K)	C_v (kJ/kg K)	C_p (kJ/kg K)	Sound speed (m/s)
0.1 MPa								
63.17	0.1	867.29	−150.74	−150.63	2.4258	1.1764	1.9998	995.61
70	0.1	838.64	−137.04	−136.92	2.6317	1.1298	2.014	926.18
77.244	0.1	806.59	−122.37	−122.25	2.8312	1.0847	2.041	852.53
77.244	0.1	4.5565	55.126	77.073	5.4116	0.77104	1.1231	174.74
80	0.1	4.3794	57.319	80.153	5.4508	0.76665	1.112	178.31
90	0.1	3.8468	65.131	91.127	5.5801	0.75661	1.0857	190.46
100	0.1	3.4366	72.807	101.91	5.6937	0.7514	1.0713	201.64
110	0.1	3.1089	80.407	112.57	5.7954	0.74844	1.0627	212.1
120	0.1	2.8401	87.959	123.17	5.8876	0.74664	1.057	221.97
130	0.1	2.6152	95.481	133.72	5.972	0.74548	1.0532	231.37
140	0.1	2.4239	102.98	144.24	6.05	0.7447	1.0504	240.36
150	0.1	2.2592	110.47	154.73	6.1224	0.74416	1.0484	249
160	0.1	2.1157	117.94	165.21	6.19	0.74378	1.0469	257.33
170	0.1	1.9895	125.41	175.67	6.2534	0.7435	1.0457	265.37
180	0.1	1.8777	132.87	186.12	6.3131	0.74329	1.0448	273.17
190	0.1	1.7779	140.32	196.57	6.3696	0.74314	1.044	280.74
200	0.1	1.6882	147.77	207	6.4232	0.74302	1.0434	288.11
210	0.1	1.6072	155.22	217.44	6.4741	0.74293	1.0429	295.28
220	0.1	1.5337	162.66	227.86	6.5226	0.74287	1.0425	302.28
230	0.1	1.4666	170.1	238.29	6.5689	0.74283	1.0422	309.11
240	0.1	1.4052	177.54	248.71	6.6132	0.7428	1.0419	315.79
250	0.1	1.3488	184.98	259.13	6.6558	0.7428	1.0417	322.33
260	0.1	1.2967	192.42	269.54	6.6966	0.74282	1.0415	328.74
270	0.1	1.2485	199.86	279.96	6.7359	0.74286	1.0414	335.01
280	0.1	1.2037	207.3	290.37	6.7738	0.74293	1.0413	341.17
290	0.1	1.1621	214.73	300.78	6.8103	0.74303	1.0413	347.22
300	0.1	1.1233	222.17	311.2	6.8457	0.74316	1.0413	353.16
400	0.1	0.84208	296.7	415.45	7.1456	0.74746	1.045	407.51
500	0.1	0.67358	371.99	520.45	7.3798	0.75921	1.0564	454.6
600	0.1	0.5613	448.82	626.98	7.574	0.77807	1.0751	496.27
700	0.1	0.48112	527.77	735.62	7.7414	0.8011	1.0981	533.87
800	0.1	0.42099	609.1	846.64	7.8896	0.82535	1.1223	568.44
900	0.1	0.37422	692.83	960.05	8.0232	0.84884	1.1457	600.68
1000	0.1	0.33681	778.82	1075.7	8.145	0.87052	1.1674	631.11
0.2 MPa								
63.192	0.2	867.37	−150.73	−150.5	2.4259	1.1765	1.9993	995.98
70	0.2	838.85	−137.09	−136.85	2.6311	1.1301	2.0133	926.89
80	0.2	794.12	−116.79	−116.53	2.9023	1.0692	2.0547	824.95
83.626	0.2	776.8	−109.3	−109.04	2.9939	1.0498	2.0797	786.84
83.626	0.2	8.6615	58.426	81.517	5.2726	0.7869	1.1801	178.95
90	0.2	7.9251	63.672	88.909	5.3578	0.7734	1.1423	187.34
100	0.2	7.0191	71.65	100.14	5.4762	0.76161	1.1082	199.35
110	0.2	6.3152	79.45	111.12	5.5809	0.75517	1.0886	210.36

(continued)

Table 7.6 (continued)

Temperature (K)	Pressure (MPa)	Density (kg/m^3)	Internal energy (kJ/kg)	Enthalpy (kJ/kg)	Entropy (kJ/kg K)	C_v (kJ/kg K)	C_p (kJ/kg K)	Sound speed (m/s)
120	0.2	5.7479	87.145	121.94	5.675	0.75133	1.0764	220.63
130	0.2	5.2788	94.773	132.66	5.7608	0.7489	1.0681	230.32
140	0.2	4.8834	102.36	143.31	5.8398	0.74729	1.0624	239.54
150	0.2	4.5449	109.91	153.91	5.9129	0.74618	1.0582	248.35
160	0.2	4.2515	117.44	164.48	5.9811	0.74538	1.055	256.82
170	0.2	3.9945	124.95	175.02	6.045	0.7448	1.0526	264.98
180	0.2	3.7674	132.44	185.53	6.1051	0.74437	1.0507	272.87
190	0.2	3.5651	139.93	196.03	6.1618	0.74404	1.0491	280.52
200	0.2	3.3837	147.41	206.51	6.2156	0.74379	1.0479	287.95
210	0.2	3.2202	154.88	216.99	6.2667	0.7436	1.0469	295.18
220	0.2	3.0719	162.35	227.45	6.3154	0.74345	1.046	302.22
230	0.2	2.9368	169.81	237.91	6.3619	0.74333	1.0453	309.1
240	0.2	2.8132	177.26	248.36	6.4064	0.74325	1.0447	315.81
250	0.2	2.6996	184.72	258.8	6.449	0.7432	1.0442	322.38
260	0.2	2.595	192.17	269.24	6.4899	0.74318	1.0438	328.81
270	0.2	2.4982	199.62	279.68	6.5293	0.74319	1.0435	335.12
280	0.2	2.4084	207.07	290.11	6.5673	0.74323	1.0433	341.29
290	0.2	2.3248	214.52	300.55	6.6039	0.7433	1.0431	347.36
300	0.2	2.247	221.97	310.98	6.6392	0.74341	1.043	353.31
400	0.2	1.6837	296.56	415.35	6.9395	0.74759	1.0458	407.75
500	0.2	1.3466	371.89	520.41	7.1739	0.7593	1.0569	454.86
600	0.2	1.1221	448.75	626.98	7.3681	0.77813	1.0755	496.53
700	0.2	0.96184	527.71	735.64	7.5356	0.80115	1.0983	534.13
800	0.2	0.84164	609.05	846.68	7.6838	0.82539	1.1224	568.69
900	0.2	0.74816	692.79	960.11	7.8174	0.84888	1.1459	600.93
1000	0.2	0.67338	778.78	1075.8	7.9393	0.87055	1.1675	631.34
0.5 MPa								
63.258	0.5	867.62	−150.7	−150.13	2.4264	1.1768	1.9978	997.1
70	0.5	839.47	−137.21	−136.62	2.6293	1.1309	2.011	929.01
80	0.5	794.96	−116.96	−116.33	2.9001	1.07	2.0508	827.73
90	0.5	745.59	−96.12	−95.449	3.146	1.0199	2.1372	720.79
93.995	0.5	723.8	−87.491	−86.8	3.24	1.0033	2.196	674.86
93.995	0.5	20.646	62.305	86.523	5.084	0.8242	1.3405	182.82
100	0.5	18.858	67.786	94.3	5.1642	0.79971	1.2577	191.75
110	0.5	16.613	76.369	106.47	5.2802	0.77827	1.1838	204.82
120	0.5	14.923	84.578	118.08	5.3813	0.76676	1.1432	216.46
130	0.5	13.584	92.573	129.38	5.4717	0.75985	1.118	227.11
140	0.5	12.488	100.43	140.47	5.5539	0.7554	1.1011	237.04
150	0.5	11.568	108.2	151.42	5.6295	0.7524	1.0892	246.41
160	0.5	10.783	115.9	162.26	5.6995	0.75029	1.0805	255.31
170	0.5	10.104	123.55	173.03	5.7648	0.74877	1.0739	263.82
180	0.5	9.509	131.17	183.75	5.826	0.74763	1.0688	272
190	0.5	8.9829	138.75	194.41	5.8837	0.74677	1.0647	279.88
200	0.5	8.514	146.32	205.04	5.9382	0.74611	1.0614	287.51

(continued)

Table 7.6 (continued)

Temperature (K)	Pressure (MPa)	Density (kg/m^3)	Internal energy (kJ/kg)	Enthalpy (kJ/kg)	Entropy (kJ/kg K)	C_v (kJ/kg K)	C_p (kJ/kg K)	Sound speed (m/s)
210	0.5	8.0[illegible]32	153.86	215.64	5.9899	0.74559	1.0588	294.9
220	0.5	7.7131	161.4	226.22	6.0391	0.74518	1.0566	302.08
230	0.5	7.3668	168.92	236.78	6.0861	0.74485	1.0547	309.08
240	0.5	7.0532	176.43	247.32	6.1309	0.7446	1.0532	315.9
250	0.5	6.7647	183.93	257.84	6.1739	0.7444	1.0518	322.56
260	0.5	6.4993	191.42	268.35	6.2151	0.74425	1.0507	329.07
270	0.5	6.2543	198.91	278.86	6.2547	0.74416	1.0498	335.43
280	0.5	6.0274	206.39	289.35	6.2929	0.74411	1.049	341.67
290	0.5	5.8166	213.87	299.84	6.3297	0.74411	1.0483	347.79
300	0.5	5.6202	221.35	310.32	6.3652	0.74415	1.0478	353.79
400	0.5	4.2057	296.15	415.03	6.6665	0.74797	1.0482	408.47
500	0.5	3.3625	371.59	520.29	6.9013	0.75955	1.0584	455.65
600	0.5	2.8017	448.51	626.97	7.0958	0.77832	1.0764	497.33
700	0.5	2.4016	527.53	735.72	7.2634	0.8013	1.099	534.91
800	0.5	2.1016	608.9	846.81	7.4117	0.82552	1.1229	569.44
900	0.5	1.8683	692.67	960.28	7.5453	0.84898	1.1462	601.65
1000	0.5	1.68[illegible]7	778.68	1076	7.6672	0.87064	1.1678	632.04
1.0 MPa								
63.368	1	868.02	−150.65	−149.5	2.4272	1.1774	1.9953	998.95
70	1	840.49	−137.42	−136.23	2.6263	1.1322	2.0074	932.51
80	1	796.35	−117.25	−116	2.8964	1.0713	2.0445	832.3
90	1	747.58	−96.54	−95.202	3.1413	1.021	2.1249	727.06
100	1	690.77	−74.631	−73.183	3.3731	0.98331	2.3048	609.42
103.75	1	665.83	−65.829	−64.327	3.4601	0.97387	2.4314	559.22
103.75	1	41.331	63.539	87.734	4.9257	0.87859	1.6518	182.79
110	1	36.932	70.254	97.331	5.0156	0.83214	1.4452	194.09
120	1	32.096	79.786	110.94	5.1342	0.79805	1.2969	208.9
130	1	28.653	88.602	123.5	5.2347	0.78057	1.2222	221.53
140	1	26.005	97.031	135.48	5.3235	0.77014	1.1774	232.83
150	1	23.876	105.22	147.1	5.4037	0.76339	1.148	243.2
160	1	22.111	113.25	158.47	5.4771	0.7588	1.1274	252.87
170	1	20.616	121.16	169.67	5.545	0.75554	1.1123	261.98
180	1	19.328	128.99	180.73	5.6082	0.75315	1.1009	270.63
190	1	18.203	136.76	191.69	5.6675	0.75135	1.092	278.91
200	1	17.212	144.48	202.58	5.7233	0.74997	1.0849	286.86
210	1	16.329	152.16	213.4	5.7761	0.7489	1.0792	294.53
220	1	15.537	159.8	224.17	5.8262	0.74804	1.0746	301.94
230	1	14.822	167.42	234.89	5.8739	0.74736	1.0707	309.13
240	1	14.173	175.02	245.58	5.9194	0.74681	1.0674	316.11
250	1	13.58	182.6	256.24	5.9629	0.74638	1.0647	322.92
260	1	13.037	190.17	266.88	6.0046	0.74603	1.0623	329.55
270	1	12.537	197.72	277.49	6.0447	0.74576	1.0603	336.03
280	1	12.075	205.27	288.08	6.0832	0.74557	1.0586	342.36
290	1	11.647	212.8	298.66	6.1203	0.74544	1.0572	348.56

(continued)

Table 7.6 (continued)

Temperature (K)	Pressure (MPa)	Density (kg/m^3)	Internal energy (kJ/kg)	Enthalpy (kJ/kg)	Entropy (kJ/kg K)	C_v (kJ/kg K)	C_p (kJ/kg K)	Sound speed (m/s)
300	1	11.249	220.33	309.23	6.1561	0.74538	1.0559	354.63
400	1	8.3992	295.46	414.52	6.4591	0.74862	1.0523	409.69
500	1	6.7114	371.09	520.09	6.6946	0.75997	1.0608	456.98
600	1	5.5916	448.13	626.97	6.8894	0.77863	1.078	498.66
700	1	4.7933	527.22	735.85	7.0572	0.80154	1.1001	536.21
800	1	4.195	608.66	847.04	7.2056	0.82572	1.1238	570.7
900	1	3.7298	692.46	960.58	7.3393	0.84916	1.1469	602.87
1000	1	3.3576	778.52	1076.3	7.4613	0.8708	1.1683	633.21
2.0 MPa								
63.587	2	868.82	−150.56	−148.25	2.4287	1.1785	1.9904	1002.6
70	2	842.51	−137.84	−135.46	2.6203	1.1348	2.0003	939.39
80	2	799.06	−117.82	−115.32	2.8893	1.0738	2.0325	841.21
90	2	751.43	−97.35	−94.688	3.1322	1.0233	2.1022	739.11
100	2	696.82	−75.888	−73.018	3.3604	0.98433	2.2518	627.3
110	2	627.79	−52.099	−48.913	3.5898	0.96374	2.6418	493.43
115.6	2	572.93	−36.065	−32.574	3.7346	0.97638	3.3302	393.23
115.6	2	90.939	59.243	81.236	4.7191	0.99942	2.8779	177.27
120	2	79.522	66.657	91.807	4.8089	0.90866	2.0851	189.73
130	2	65.485	79.043	109.58	4.9515	0.83561	1.5791	209.29
140	2	57.11	89.329	124.35	5.061	0.80511	1.3965	224.25
150	2	51.18	98.712	137.79	5.1538	0.78793	1.3003	236.99
160	2	46.634	107.59	150.48	5.2356	0.77705	1.2411	248.33
170	2	42.981	116.14	162.68	5.3096	0.76968	1.2013	258.7
180	2	39.951	124.48	174.54	5.3774	0.76446	1.1729	268.33
190	2	37.382	132.65	186.16	5.4403	0.76062	1.1517	277.38
200	2	35.164	140.71	197.59	5.4989	0.75772	1.1354	285.96
210	2	33.224	148.68	208.88	5.554	0.75549	1.1226	294.14
220	2	31.508	156.57	220.05	5.606	0.75372	1.1122	301.98
230	2	29.976	164.41	231.13	5.6552	0.75232	1.1038	309.53
240	2	28.599	172.2	242.13	5.702	0.75119	1.0968	316.83
250	2	27.352	179.95	253.07	5.7467	0.75027	1.0909	323.9
260	2	26.216	187.66	263.95	5.7894	0.74952	1.0859	330.76
270	2	25.177	195.35	274.79	5.8303	0.74892	1.0816	337.43
280	2	24.222	203.01	285.58	5.8695	0.74844	1.078	343.93
290	2	23.34	210.66	296.35	5.9073	0.74806	1.0748	350.28
300	2	22.523	218.29	307.08	5.9437	0.74779	1.0721	356.48
400	2	16.747	294.09	413.52	6.25	0.74989	1.0603	412.22
500	2	13.368	370.09	519.7	6.4869	0.76081	1.0656	459.67
600	2	11.135	447.36	626.97	6.6824	0.77925	1.0811	501.35
700	2	9.5467	526.61	736.11	6.8506	0.80204	1.1023	538.83
800	2	8.357	608.17	847.49	6.9993	0.82613	1.1254	573.23
900	2	7.4321	692.07	961.17	7.1332	0.84951	1.1481	605.3
1000	2	6.6921	778.19	1077	7.2552	0.8711	1.1692	635.55
5.0 MPa								

(continued)

Table 7.6 (continued)

Temperature (K)	Pressure (MPa)	Density (kg/m^3)	Internal energy (kJ/kg)	Enthalpy (kJ/kg)	Entropy (kJ/kg K)	C_v (kJ/kg K)	C_p (kJ/kg K)	Sound speed (m/s)
64.242	5	871.19	−150.25	−144.51	2.4333	1.1817	1.9764	1013.4
70	5	848.33	−139.01	−133.12	2.6031	1.1424	1.9812	959.16
80	5	806.76	−119.42	−113.22	2.8688	1.0812	2.0013	866.36
90	5	762.03	−99.573	−93.012	3.1068	1.0302	2.0466	772.11
100	5	712.56	−79.161	−72.144	3.3265	0.98927	2.1371	673.24
110	5	655.11	−57.6	−49.967	3.5378	0.95944	2.3211	565.77
120	5	581.7	−33.513	−24.917	3.7555	0.94717	2.7623	442.95
130	5	460.36	−1.5661	9.2951	4.0284	0.98996	4.6592	288.81
140	5	234.95	49.573	70.854	4.4847	0.98158	4.4648	208.84
150	5	168.9	72.421	102.02	4.7005	0.88172	2.3653	226.09
160	5	140.44	86.997	122.6	4.8335	0.83916	1.8337	241.41
170	5	122.91	98.931	139.61	4.9367	0.81487	1.5949	254.67
180	5	110.48	109.57	154.83	5.0237	0.79916	1.4596	266.49
190	5	100.98	119.45	168.96	5.1002	0.7883	1.373	277.24
200	5	93.366	128.82	182.38	5.169	0.78044	1.313	287.17
210	5	87.069	137.85	195.28	5.2319	0.77456	1.2692	296.46
220	5	81.734	146.62	207.79	5.2902	0.77003	1.2359	305.21
230	5	77.133	155.2	220.02	5.3445	0.76647	1.2099	313.52
240	5	73.107	163.62	232.01	5.3955	0.76362	1.1891	321.46
250	5	69.544	171.92	243.81	5.4437	0.76132	1.1721	329.07
260	5	66.36	180.12	255.46	5.4894	0.75943	1.1581	336.4
270	5	63.492	188.23	266.98	5.5329	0.75787	1.1463	343.47
280	5	60.892	196.28	278.39	5.5744	0.75658	1.1363	350.32
290	5	58.52	204.27	289.71	5.6141	0.75552	1.1277	356.96
300	5	56.344	212.21	300.95	5.6522	0.75466	1.1203	363.42
400	5	41.44	290.07	410.73	5.9683	0.75358	1.0836	420.44
500	5	32.993	367.15	518.7	6.2092	0.76328	1.0793	468.09
600	5	27.477	445.11	627.08	6.4068	0.78109	1.0901	509.6
700	5	23.569	524.83	736.98	6.5762	0.8035	1.1086	546.8
800	5	20.647	606.73	848.9	6.7256	0.82735	1.13	580.88
900	5	18.376	690.89	962.99	6.8599	0.85055	1.1516	612.63
1000	5	16.559	777.22	1079.2	6.9823	0.87201	1.172	642.58
10.0 MPa								
65.321	10	875.01	−149.72	−138.29	2.4409	1.1865	1.955	1030.7
70	10	857.35	−140.81	−129.14	2.5762	1.1546	1.9546	989.62
80	10	818.38	−121.8	−109.58	2.8374	1.0933	1.9604	904.07
90	10	777.36	−102.75	−89.886	3.0693	1.0419	1.9806	819.44
100	10	733.61	−83.525	−69.894	3.2799	0.99968	2.0221	734.22
110	10	686.03	−63.915	−49.338	3.4758	0.96584	2.0955	647.61
120	10	632.9	−43.622	−27.821	3.6629	0.94014	2.2178	559.64
130	10	571.71	−22.235	−4.7433	3.8476	0.92277	2.4115	472.34
140	10	499.8	0.65178	20.66	4.0357	0.91267	2.676	391.97
150	10	419.17	24.663	48.52	4.2278	0.9031	2.8538	331.25
160	10	344.17	47.384	76.439	4.4081	0.8849	2.668	300.78

(continued)

Table 7.6 (continued)

Temperature (K)	Pressure (MPa)	Density (kg/m^3)	Internal energy (kJ/kg)	Enthalpy (kJ/kg)	Entropy (kJ/kg K)	C_v (kJ/kg K)	C_p (kJ/kg K)	Sound speed (m/s)
170	10	287.98	66.547	101.27	4.5587	0.86173	2.2975	292.7
180	10	248.74	82.426	122.63	4.6809	0.8407	1.9905	294.66
190	10	220.6	96.063	141.39	4.7824	0.82383	1.7763	300.47
200	10	199.44	108.23	158.37	4.8695	0.81074	1.6281	307.71
210	10	182.86	119.41	174.1	4.9463	0.8006	1.5225	315.42
220	10	169.43	129.89	188.92	5.0152	0.79265	1.4448	323.21
230	10	158.26	139.87	203.05	5.0781	0.78635	1.3857	330.91
240	10	148.78	149.46	216.67	5.136	0.78128	1.3396	338.43
250	10	140.59	158.75	229.88	5.19	0.77714	1.3028	345.77
260	10	133.43	167.81	242.75	5.2405	0.77373	1.2729	352.89
270	10	127.1	176.67	255.35	5.288	0.7709	1.2481	359.83
280	10	121.44	185.38	267.73	5.333	0.76853	1.2274	366.57
290	10	116.34	193.96	279.91	5.3758	0.76654	1.2099	373.13
300	10	111.73	202.43	291.93	5.4165	0.76486	1.1949	379.52
400	10	81.186	283.65	406.82	5.7476	0.75934	1.1189	436.03
500	10	64.498	362.47	517.51	5.9946	0.76723	1.1002	483.06
600	10	53.738	441.51	627.6	6.1953	0.78409	1.104	523.87
700	10	46.151	521.98	738.66	6.3665	0.80591	1.1184	560.38
800	10	40.486	604.43	851.42	6.5171	0.82936	1.1373	593.81
900	10	36.083	689.01	966.15	6.6522	0.85227	1.1572	624.96
1000	10	32.556	775.67	1082.8	6.7751	0.87352	1.1763	654.37
50.0 MPa								
73.495	50	901.56	−144.62	−89.162	2.5003	1.2086	1.8388	1150.9
80	50	883.17	−133.87	−77.257	2.6555	1.1692	1.8216	1114.3
90	50	855	−117.65	−59.168	2.8686	1.1172	1.7965	1061
100	50	827.06	−101.78	−41.321	3.0566	1.074	1.773	1010.9
110	50	799.44	−86.247	−23.704	3.2246	1.0377	1.7508	963.85
120	50	772.22	−71.05	−6.3012	3.376	1.007	1.7299	919.79
130	50	745.44	−56.176	10.898	3.5137	0.98081	1.71	878.83
140	50	719.18	−41.621	27.902	3.6397	0.95833	1.6909	841.08
150	50	693.5	−27.38	44.718	3.7557	0.9389	1.6723	806.6
160	50	668.46	−13.451	61.348	3.8631	0.92202	1.6537	775.41
170	50	644.15	0.1685	77.79	3.9628	0.90727	1.6348	747.47
180	50	620.63	13.478	94.042	4.0557	0.89432	1.6154	722.72
190	50	597.97	26.48	110.1	4.1425	0.88289	1.5953	701
200	50	576.24	39.177	125.95	4.2238	0.87277	1.5747	682.14
210	50	555.47	51.574	141.59	4.3001	0.86375	1.5535	665.92
220	50	535.7	63.678	157.02	4.3719	0.8557	1.532	652.1
230	50	516.93	75.501	172.23	4.4395	0.84848	1.5103	640.44
240	50	499.16	87.053	187.22	4.5033	0.84198	1.4888	630.7
250	50	482.36	98.348	202	4.5636	0.83612	1.4677	622.64
260	50	466.52	109.4	216.58	4.6208	0.83083	1.4471	616.05
270	50	451.58	120.22	230.95	4.675	0.82605	1.4272	610.73
280	50	437.5	130.84	245.12	4.7266	0.82171	1.4082	606.51

(continued)

Table 7.6 (continued)

Temperature (K)	Pressure (MPa)	Density (kg/m^3)	Internal energy (kJ/kg)	Enthalpy (kJ/kg)	Entropy (kJ/kg K)	C_v (kJ/kg K)	C_p (kJ/kg K)	Sound speed (m/s)
290	50	424.23	141.25	259.11	4.7757	0.81779	1.39	603.25
300	50	411.72	151.49	272.93	4.8225	0.81424	1.3728	600.81
400	50	318.73	246.48	403.35	5.1986	0.79422	1.2498	602.98
500	50	261.74	334.05	525.08	5.4704	0.79361	1.1928	624.19
600	50	223.19	419.06	643.09	5.6857	0.80515	1.1718	649.42
700	50	195.17	503.87	760.05	5.866	0.82338	1.1698	675.08
800	50	173.77	589.62	877.36	6.0226	0.84424	1.1773	700.29
900	50	156.82	676.81	995.65	6.1619	0.86522	1.189	724.89
1000	50	143.01	765.59	1115.2	6.2879	0.88496	1.2022	748.86
100.0 MPa								
82.799	100	928.56	−137.2	−29.507	2.5648	1.218	1.7586	1272.6
90	100	912.52	−126.5	−16.917	2.7106	1.1815	1.7384	1244.6
100	100	890.63	−111.95	0.32923	2.8923	1.1382	1.7111	1207.7
110	100	869.26	−97.733	17.307	3.0541	1.1018	1.6846	1173.1
120	100	848.46	−83.836	34.024	3.1996	1.0707	1.6589	1140.6
130	100	828.27	−70.245	50.489	3.3314	1.0438	1.6342	1110.1
140	100	808.68	−56.947	66.712	3.4517	1.0205	1.6106	1081.7
150	100	789.7	−43.926	82.704	3.562	1	1.588	1055.3
160	100	771.33	−31.169	98.477	3.6638	0.98194	1.5666	1030.8
170	100	753.56	−18.663	114.04	3.7582	0.9659	1.5464	1008.1
180	100	736.37	−6.3933	129.41	3.846	0.95159	1.5271	987.19
190	100	719.76	5.6517	144.59	3.9281	0.93876	1.5089	967.97
200	100	703.71	17.484	159.59	4.005	0.92722	1.4916	950.34
210	100	688.2	29.116	174.42	4.0774	0.9168	1.4752	934.2
220	100	673.23	40.556	189.09	4.1457	0.90735	1.4595	919.46
230	100	658.78	51.817	203.61	4.2102	0.89877	1.4446	906.03
240	100	644.83	62.907	217.99	4.2714	0.89094	1.4303	893.82
250	100	631.37	73.835	232.22	4.3295	0.88379	1.4167	882.74
260	100	618.38	84.609	246.32	4.3848	0.87725	1.4036	872.7
270	100	605.85	95.239	260.3	4.4375	0.87125	1.3911	863.64
280	100	593.77	105.73	274.15	4.4879	0.86574	1.3792	855.46
290	100	582.12	116.1	287.88	4.5361	0.86069	1.3677	848.09
300	100	570.89	126.34	301.5	4.5823	0.85604	1.3568	841.48
400	100	478.03	223.39	432.59	4.96	0.82692	1.2719	804.98
500	100	411.64	314.2	557.13	5.2381	0.82	1.2243	798.27
600	100	362.27	402.28	678.31	5.4591	0.8271	1.2029	803.96
700	100	324.15	489.76	798.26	5.644	0.8421	1.198	815.34
800	100	293.75	577.78	918.21	5.8042	0.86053	1.2021	829.64
900	100	268.89	666.91	1038.8	5.9462	0.87961	1.2106	845.59
1000	100	248.12	757.35	1160.4	6.0743	0.89783	1.2208	862.5

Reference

1. R. Span, E.W. Lemmon, R.T. Jacobsen, W. Wagner, A. Yokozeki, A reference equation of state for the thermodynamic properties of nitrogen for temperatures from 63.151 to 1000 K and pressures to 2200 MPa. J. Phys. Chem. Ref. Data. **29**(6), 1361–1433 (2000)

Chapter 8
Air

Although air is a mixture of several pure cryogenic fluids, convenience led to the development of a pseudo-pure fluid formulation in addition to a mixture model for nitrogen, argon, and oxygen. This formulation is analogous to other pure fluid formulations in this book, except as noted below. The pseudo-pure fluid formulation is valid for single-phase states only, including states along the saturation line, but not within the 2-phase region. Because the fugacities, which are required to determine vapor-liquid equilibrium for mixtures, are a function of composition, the equation of state cannot be used to determine the bubble or dew-point pressures. Thus, the pressure ancillaries given below are not just for convenience, but are required to identify the saturation states. Two ancillaries are given for the saturated pressure, one for the bubble-point pressure (the saturation state in the liquid phase) and one for the dew-point pressure (the saturation state in the vapor phase). The behavior of mixtures is quite different from that of pure fluids in that the pressure (and temperature) changes as a function of quality from liquid to vapor, and thus a single vapor-pressure equation cannot be used. To obtain the densities at saturation, the ancillary equations for density can be used as an initial estimate, and then an iterative routine must be used to locate the correct density from the equation of state as a function of the temperature and the corresponding pressure calculated from either the bubble or dew-point pressure ancillary equations.

8.1 Formulation for Air

The formulation for air was developed by Lemmon et al. [1]. The uncertainties in the formulation are as follows.

In the range from the solidification point to 873 K at pressures to 70 MPa, the estimated uncertainty of density values calculated with the equation of state is 0.1%. The estimated uncertainty of calculated speed of sound values is 0.2% and

J.W. Leachman et al., *Thermodynamic Properties of Cryogenic Fluids*,
International Cryogenics Monograph Series, DOI 10.1007/978-3-319-57835-4_8

that for calculated heat capacities is 1%. At temperatures above 873 K and 70 MPa, the estimated uncertainty of calculated density values is 0.5% increasing to 1.0% at 2000 K and 2000 MPa.

An additional term must be added to Eq. 2.23 to account for the entropy of mixing ($S_{mix} = \Sigma_i x_i \ln x_i = -0.563543$) (Tables 8.1, 8.2, 8.3 and 8.4).

Table 8.1 Fixed point properties and correlation limits for air

	Temperature (K)	Pressure (kPa)	Density (mol/dm^3)
Critical point	132.6312	3785.02	10.4477
Triple point	59.75	5.2646	33.067
Upper limit	2,000	2,000,000	53.73
Lower limit	59.75		

Table 8.2 Fluid constants and normal boiling point for air

Molar mass	28.96546 g/mol
Ideal-gas constant	8.31451 J/(mol K)
Pitzer's acentric factor	0.0335
Normal boiling point	78.903 K

Table 8.3 Fundamental equation of state coefficients and exponents for air

i	N_i	t_i	d_i	p_i
1	0.118160747	0	1	
2	0.713116392	0.33	1	
3	−1.618241921	1.01	1	
4	0.071414018	0	2	
5	−0.08654214	0	3	
6	0.134211177	0.15	3	
7	0.01126267	0	4	
8	−0.042053323	0.2	4	
9	0.034900843	0.35	4	
10	0.000164957	1.35	6	
11	−0.101365038	1.6	1	1
12	−0.173813691	0.8	3	1
13	−0.047210318	0.95	5	1
14	−0.012252355	1.25	6	1
15	−0.14662961	3.6	1	2
16	−0.031605588	6	3	2
17	0.000233595	3.25	11	2
18	0.014828789	3.5	1	3
19	−0.009387829	15	3	3

Table 8.4 Ancillary function coefficients and exponents for air

	α^0	
i	a_i	k_i
0	3.490888	
1	−3.4666226577	
2	3.3037934171	
3	2.3955256×10^{-6}	1
4	7.1721112×10^{-9}	2
5	$-3.1154131 \times 10^{-13}$	3
6	0.22380669	−1.5
7	0.79130951	3364.011
8	0.21223677	2242.45

$m = 6$, $n = 8$ (there are no polynomial terms in the first summation)

	Vapor pressure (dew point)		Liquid pressure (bubble point)		Saturated vapor density (dew point)		Saturated liquid density (bubble point)	
i	N_i	k_i	N_i	k_i	N_i	k_i	N_i	k_i
1	−0.1567266	0.5	0.2260724	0.5	−2.0466	0.41	9.72015	0.55
2	−5.539635	1	−7.080499	1	−4.752	1	−43.4531	1.0
3	0.7567212	2.5	5.700283	1.5	−13.259	2.8	103.196	1.4
4	−3.514322	4	−12.44017	2	−47.652	6.5	−145.477	1.9
5			17.81926	2.5			112.618	2.4
6			−10.81364	3			−34.3921	3.0

8.2 Property Plots for Air

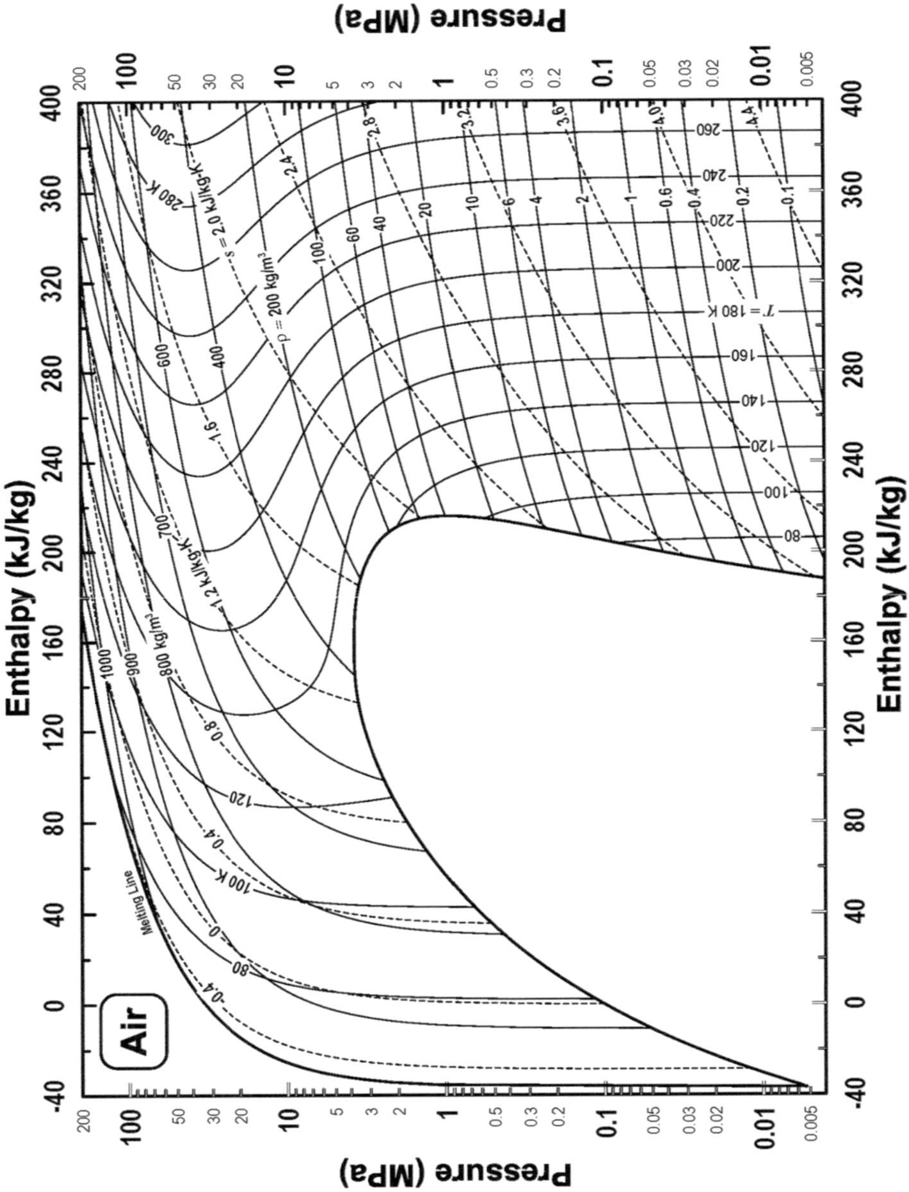
Air
Enthalpy (kJ/kg)
Pressure (MPa)
Melting Line
T = 180 K
s = 2.0 kJ/kg-K
1.2 kJ/kg-K
ρ = 200 kg/m³
800 kg/m³
100 K
280 K

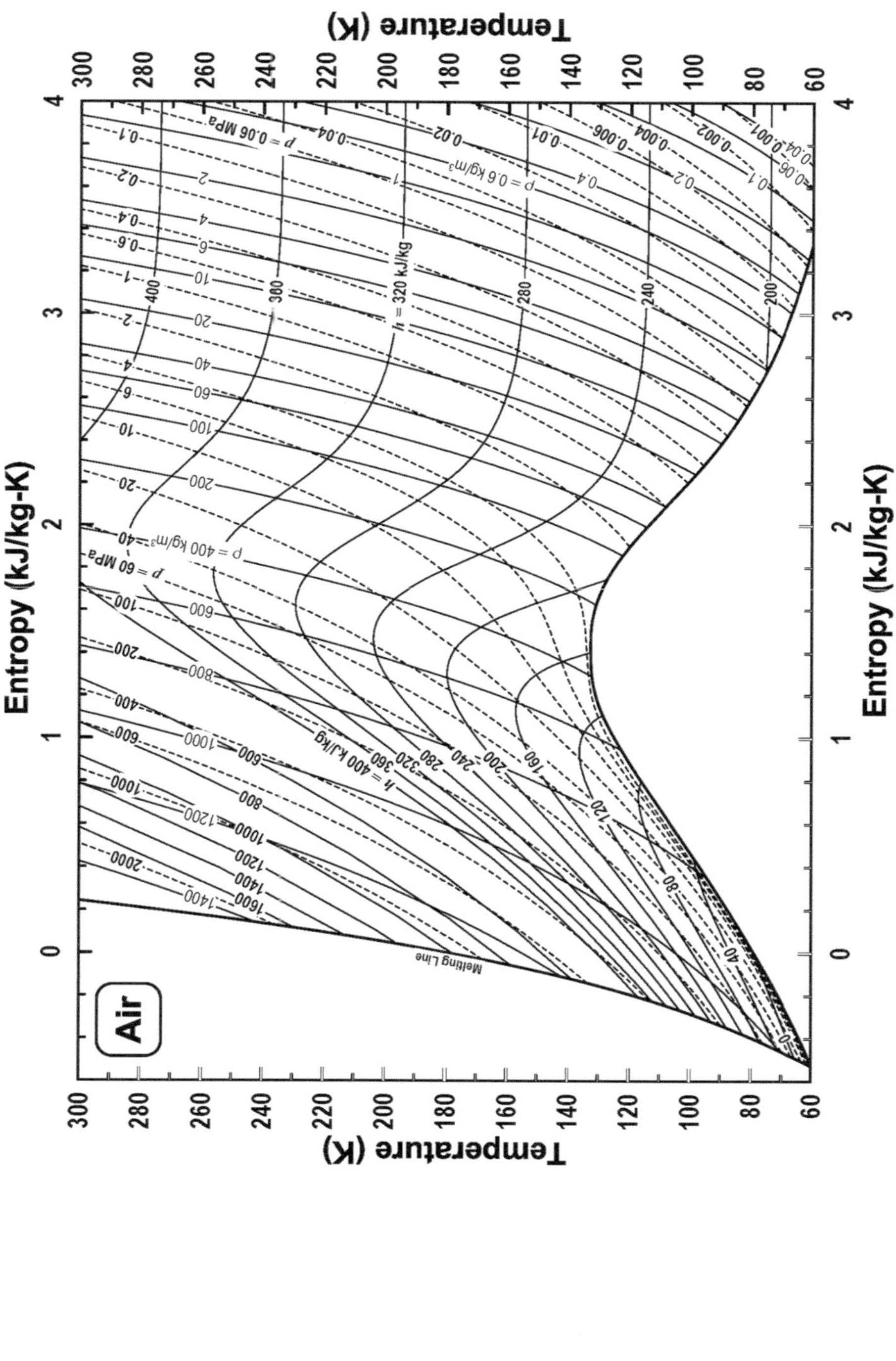
Air
Entropy (kJ/kg-K)
Temperature (K)
Melting Line

8.3 Thermodynamic Data Tables for Air

See Tables 8.5 and 8.6.

Table 8.5 Thermodynamic properties of saturated air

Liquid temp. (K)	Vapor temp. (K)	Liquid pressure (MPa)	Vapor pressure (MPa)	Liquid density (kg/m^3)	Vapor density (kg/m^3)	Liquid enthalpy (kJ/kg)	Vapor enthalpy (kJ/kg)	Liquid entropy (kJ/kg K)	Vapor entropy (kJ/kg K)	Liquid C_v (kJ/kg K)	Vapor C_v (kJ/kg K)	Liquid sound speed (m/s)	Vapor sound speed (m/s)
59.75	59.75	0.005265	0.002432	957.79	0.14213	−36.655	185.5	−0.53042	3.3387	1.1742	0.71826	1030.2	154.81
64	64	0.012	0.006325	940.12	0.34609	−28.57	189.59	−0.39983	3.1315	1.1423	0.72108	995.65	159.97
68	68	0.02352	0.013713	923.2	0.70909	−20.949	193.33	−0.28451	2.9676	1.1149	0.72501	962.13	164.51
72	72	0.042498	0.026921	905.95	1.3227	−13.303	196.92	−0.17554	2.8277	1.0896	0.73037	927.59	168.68
76	76	0.071786	0.048702	888.3	2.2859	−5.6153	200.33	−0.07208	2.7069	1.0663	0.7373	892	172.47
78.903	81.72	0.10133	0.10133	875.21	4.4974	0	204.81	0	2.5598	1.0505	0.75007	865.5	177.13
80	80	0.11462	0.082321	870.18	3.7115	2.1318	203.52	0.02664	2.6013	1.0448	0.74587	855.33	175.83
84	84	0.17453	0.13147	851.5	5.7249	9.9598	206.44	0.12127	2.508	1.0249	0.75611	817.51	178.73
88	88	0.25531	0.20018	832.15	8.4659	17.893	209.05	0.21242	2.4244	1.0067	0.76802	778.46	181.15
92	92	0.36091	0.29273	811.99	12.092	25.962	211.31	0.30065	2.3485	0.99001	0.78156	738.11	183.06
96	96	0.49543	0.41359	790.86	16.786	34.2	213.17	0.38651	2.2784	0.975	0.79675	696.38	184.44
100	100	0.66313	0.56742	768.55	22.769	42.652	214.58	0.47056	2.2128	0.96171	0.81365	653.18	185.28
104	104	0.86836	0.75903	744.79	30.319	51.37	215.46	0.55338	2.1501	0.95028	0.83245	608.43	185.55
108	108	1.1156	0.9934	719.23	39.803	60.425	215.7	0.63561	2.0889	0.941	0.85351	562.02	185.22
112	112	1.4095	1.2757	691.36	51.738	69.91	215.19	0.71805	2.0277	0.9343	0.87752	513.8	184.28
116	116	1.7546	1.6115	660.45	66.901	79.962	213.71	0.80174	1.9647	0.93095	0.90564	463.43	182.69
120	120	2.1557	2.0067	625.29	86.574	90.804	210.93	0.88832	1.8972	0.93249	0.93994	410.18	180.43
124	124	2.6173	2.4682	583.47	113.19	102.87	206.25	0.98091	1.8206	0.9425	0.98448	352.26	177.49
128	128	3.1431	3.0055	528.4	152.64	117.3	198.22	1.0879	1.7245	0.97257	1.0485	284.97	173.94
132.53	132.53	3.7887	3.7529	350.51	270.9	154.43	171.57	1.3608	1.491				

Table 8.6 Thermodynamic properties of air

Temperature (K)	Pressure (MPa)	Density (kg/m^3)	Internal energy (kJ/kg)	Enthalpy (kJ/kg)	Entropy (kJ/kg K)	C_v (kJ/kg K)	C_p (kJ/kg K)	Sound speed (m/s)
0.1 MPa								
59.767	0.1	957.87	−36.654	−36.549	−0.53031	1.1742	1.9007	1030.5
60	0.1	956.9	−36.211	−36.106	−0.52291	1.1724	1.9007	1028.6
70	0.1	914.75	−17.19	−17.08	−0.22964	1.1021	1.9076	945.41
78.788	0.1	875.73	−0.33781	−0.22362	−0.00282	1.0511	1.9324	866.57
81.609	0.1	4.443	182.22	204.72	2.5625	0.74979	1.0888	177.05
90	0.1	3.9854	188.63	213.72	2.6674	0.73609	1.0587	187.14
100	0.1	3.5578	196.1	224.2	2.7779	0.72802	1.0397	198.22
110	0.1	3.2173	203.46	234.54	2.8764	0.72359	1.0288	208.55
120	0.1	2.9385	210.76	244.79	2.9656	0.72096	1.0219	218.29
130	0.1	2.7054	218.02	254.99	3.0472	0.71931	1.0174	227.56
140	0.1	2.5073	225.26	265.15	3.1225	0.71823	1.0142	236.42
150	0.1	2.3368	232.48	275.28	3.1924	0.71751	1.012	244.92
160	0.1	2.1883	239.69	285.39	3.2577	0.71701	1.0103	253.12
170	0.1	2.0577	246.89	295.48	3.3189	0.71666	1.0091	261.04
180	0.1	1.942	254.08	305.57	3.3765	0.71642	1.0081	268.72
190	0.1	1.8387	261.26	315.65	3.431	0.71625	1.0073	276.17
200	0.1	1.746	268.44	325.72	3.4827	0.71615	1.0067	283.42
210	0.1	1.6622	275.62	335.78	3.5318	0.7161	1.0063	290.47
220	0.1	1.5861	282.8	345.84	3.5786	0.71611	1.006	297.36
230	0.1	1.5167	289.97	355.9	3.6233	0.71616	1.0057	304.08
240	0.1	1.4532	297.14	365.96	3.6661	0.71625	1.0056	310.64
250	0.1	1.3948	304.32	376.01	3.7071	0.7164	1.0055	317.07
260	0.1	1.3409	311.49	386.07	3.7466	0.7166	1.0055	323.36
270	0.1	1.2911	318.67	396.12	3.7845	0.71685	1.0056	329.52
280	0.1	1.2448	325.85	406.18	3.8211	0.71716	1.0058	335.57
290	0.1	1.2018	333.03	416.24	3.8564	0.71753	1.006	341.5
300	0.1	1.1616	340.21	426.3	3.8905	0.71797	1.0064	347.32
400	0.1	0.87077	412.41	527.26	4.1809	0.72641	1.0141	400.45
500	0.1	0.69652	485.82	629.4	4.4087	0.74242	1.0299	446.34
600	0.1	0.58041	561.12	733.42	4.5983	0.7639	1.0512	487.01
700	0.1	0.49749	638.71	839.72	4.7621	0.78775	1.075	523.83
800	0.1	0.43531	718.69	948.41	4.9072	0.81151	1.0987	557.78
900	0.1	0.38695	800.97	1059.4	5.0379	0.83376	1.1209	589.53
1000	0.1	0.34826	885.37	1172.5	5.1571	0.85384	1.141	619.53
0.2 MPa								
59.785	0.2	957.94	−36.647	−36.438	−0.53019	1.1742	1.9004	1030.9
60	0.2	957.06	−36.238	−36.029	−0.52336	1.1726	1.9003	1029.1
70	0.2	914.95	−17.226	−17.007	−0.23016	1.1023	1.9071	946.03
80	0.2	870.4	1.9584	2.1882	0.026119	1.0449	1.9368	856.02
85.388	0.2	844.87	12.462	12.699	0.15326	1.0184	1.967	804.11
87.991	0.2	8.4587	185.4	209.04	2.4246	0.76799	1.1441	181.14
90	0.2	8.2298	187.02	211.33	2.4502	0.7613	1.1288	183.74
100	0.2	7.2765	194.86	222.35	2.5664	0.74186	1.0821	195.76

(continued)

Table 8.6 (continued)

Temperature (K)	Pressure (MPa)	Density (kg/m³)	Internal energy (kJ/kg)	Enthalpy (kJ/kg)	Entropy (kJ/kg K)	C_v (kJ/kg K)	C_p (kJ/kg K)	Sound speed (m/s)
110	0.2	6.5412	202.46	233.04	2.6683	0.73237	1.0577	206.69
120	0.2	5.9508	209.93	243.54	2.7597	0.72691	1.0429	216.85
130	0.2	5.4637	217.31	253.91	2.8427	0.72352	1.0333	226.43
140	0.2	5.0535	224.64	264.21	2.9191	0.72132	1.0267	235.52
150	0.2	4.7026	231.93	274.45	2.9897	0.71984	1.022	244.21
160	0.2	4.3987	239.19	284.66	3.0556	0.71882	1.0186	252.55
170	0.2	4.1325	246.43	294.83	3.1172	0.71809	1.016	260.59
180	0.2	3.8973	253.66	304.98	3.1752	0.71758	1.014	268.36
190	0.2	3.6879	260.88	315.11	3.23	0.71721	1.0124	275.89
200	0.2	3.5002	268.09	325.23	3.2819	0.71695	1.0111	283.21
210	0.2	3.3309	275.29	335.33	3.3312	0.71678	1.0102	290.32
220	0.2	3.1775	282.49	345.43	3.3782	0.71669	1.0094	297.25
230	0.2	3.0377	289.68	355.52	3.4231	0.71666	1.0088	304.01
240	0.2	2.9097	296.87	365.61	3.466	0.7167	1.0083	310.62
250	0.2	2.7922	304.06	375.69	3.5071	0.71679	1.008	317.07
260	0.2	2.684	311.25	385.77	3.5467	0.71695	1.0078	323.39
270	0.2	2.5838	318.44	395.84	3.5847	0.71717	1.0077	329.58
280	0.2	2.4909	325.63	405.92	3.6213	0.71745	1.0076	335.64
290	0.2	2.4045	332.82	416	3.6567	0.71779	1.0077	341.59
300	0.2	2.3239	340.01	426.07	3.6909	0.71821	1.0079	347.43
400	0.2	1.7412	412.28	527.14	3.9816	0.72654	1.0149	400.65
500	0.2	1.3926	485.73	629.34	4.2096	0.7425	1.0303	446.57
600	0.2	1.1604	561.05	733.4	4.3992	0.76396	1.0515	487.24
700	0.2	0.99463	638.65	839.73	4.5631	0.7878	1.0752	524.06
800	0.2	0.87032	718.64	948.44	4.7082	0.81155	1.0988	558
900	0.2	0.77365	800.93	1059.4	4.8389	0.83379	1.121	589.74
1000	0.2	0.69631	885.34	1172.6	4.9581	0.85387	1.141	619.73
0.5 MPa								
59.838	0.5	958.18	−36.625	−36.103	−0.52984	1.1743	1.8994	1031.9
60	0.5	957.52	−36.319	−35.796	−0.52471	1.173	1.8993	1030.7
70	0.5	915.53	−17.334	−16.787	−0.2317	1.1028	1.9055	947.9
80	0.5	871.16	1.8128	2.3868	0.024295	1.0455	1.934	858.41
90	0.5	822.86	21.405	22.013	0.25538	0.99847	1.9993	760.54
96.121	0.5	790.2	33.82	34.453	0.38908	0.97457	2.0706	695.09
98.358	0.5	20.142	189.24	214.06	2.2393	0.8065	1.2879	185
100	0.5	19.659	190.72	216.15	2.2604	0.79653	1.2611	187.43
110	0.5	17.265	199.25	228.21	2.3753	0.762	1.1636	200.71
120	0.5	15.486	207.29	239.58	2.4743	0.74595	1.1152	212.35
130	0.5	14.084	215.08	250.58	2.5623	0.73671	1.0862	222.95
140	0.5	12.941	222.7	261.34	2.6421	0.73088	1.0673	232.79
150	0.5	11.984	230.22	271.94	2.7153	0.72701	1.0541	242.06
160	0.5	11.168	237.66	282.43	2.783	0.72434	1.0446	250.85
170	0.5	10.462	245.05	292.84	2.8461	0.72246	1.0376	259.25
180	0.5	9.8446	252.4	303.19	2.9052	0.72109	1.0322	267.32

(continued)

Table 8.6 (continued)

Temperature (K)	Pressure (MPa)	Density (kg/m³)	Internal energy (kJ/kg)	Enthalpy (kJ/kg)	Entropy (kJ/kg K)	C_v (kJ/kg K)	C_p (kJ/kg K)	Sound speed (m/s)
190	0.5	9.299	259.72	313.49	2.9609	0.7201	1.028	275.09
200	0.5	8.8128	267.02	323.75	3.0136	0.71936	1.0246	282.61
210	0.5	8.3766	274.29	333.98	3.0635	0.71883	1.0219	289.89
220	0.5	7.9827	281.56	344.19	3.111	0.71844	1.0198	296.97
230	0.5	7.6251	288.81	354.38	3.1563	0.71818	1.018	303.85
240	0.5	7.299	296.05	364.55	3.1996	0.71803	1.0166	310.56
250	0.5	7.0001	303.29	374.71	3.241	0.71798	1.0154	317.11
260	0.5	6.7252	310.52	384.86	3.2808	0.71801	1.0145	323.51
270	0.5	6.4715	317.74	395	3.3191	0.71812	1.0138	329.76
280	0.5	6.2365	324.97	405.14	3.356	0.71831	1.0133	335.89
290	0.5	6.0182	332.19	415.27	3.3915	0.71858	1.0129	341.89
300	0.5	5.8149	339.41	425.4	3.4259	0.71893	1.0127	347.78
400	0.5	4.3506	411.88	526.8	3.7175	0.72692	1.0173	401.25
500	0.5	3.478	485.43	629.19	3.946	0.74276	1.0318	447.25
600	0.5	2.8978	560.82	733.37	4.1358	0.76415	1.0525	487.94
700	0.5	2.4838	638.47	839.77	4.2998	0.78794	1.0758	524.74
800	0.5	2.1735	718.49	948.54	4.445	0.81167	1.0993	558.67
900	0.5	1.9322	800.81	1059.6	4.5758	0.83389	1.1214	590.39
1000	0.5	1.7392	885.24	1172.7	4.695	0.85395	1.1413	620.36
1.0 MPa								
59.928	1	958.57	−36.589	−35.546	−0.52924	1.1744	1.8977	1033.7
60	1	958.27	−36.452	−35.409	−0.52695	1.1738	1.8977	1033.1
70	1	916.48	−17.512	−16.42	−0.23425	1.1037	1.9028	950.98
80	1	872.4	1.5729	2.7192	0.02128	1.0464	1.9295	862.35
90	1	824.57	21.071	22.284	0.25164	0.99933	1.9909	765.79
100	1	770.28	41.453	42.751	0.46718	0.9621	2.1182	658.17
106.22	1	730.86	54.976	56.345	0.59901	0.94485	2.2675	582.9
108.1	1	40.075	190.75	215.7	2.0873	0.85409	1.5397	185.21
110	1	38.76	192.76	218.56	2.1135	0.83585	1.4737	188.64
120	1	33.492	202.34	232.2	2.2323	0.78417	1.2843	204.02
130	1	29.82	211.03	244.56	2.3314	0.76112	1.1977	216.81
140	1	27.024	219.27	256.27	2.4181	0.74794	1.1476	228.13
150	1	24.788	227.23	267.57	2.4961	0.73954	1.1153	238.47
160	1	22.941	235.02	278.61	2.5674	0.73389	1.0929	248.07
170	1	21.38	242.68	289.45	2.6331	0.72993	1.0768	257.1
180	1	20.037	250.25	300.16	2.6943	0.72708	1.0647	265.67
190	1	18.867	257.75	310.76	2.7516	0.72499	1.0554	273.85
200	1	17.835	265.2	321.27	2.8056	0.72343	1.0482	281.7
210	1	16.918	272.61	331.72	2.8565	0.72226	1.0424	289.27
220	1	16.095	279.99	342.12	2.9049	0.72138	1.0377	296.58
230	1	15.352	287.34	352.48	2.951	0.72073	1.0338	303.67
240	1	14.679	294.68	362.8	2.9949	0.72027	1.0307	310.55
250	1	14.064	301.99	373.1	3.0369	0.71995	1.0281	317.24
260	1	13.5	309.29	383.37	3.0772	0.71977	1.0259	323.77

(continued)

Table 8.6 (continued)

Temperature (K)	Pressure (MPa)	Density (kg/m^3)	Internal energy (kJ/kg)	Enthalpy (kJ/kg)	Entropy (kJ/kg K)	C_v (kJ/kg K)	C_p (kJ/kg K)	Sound speed (m/s)
270	1	12.981	316.58	393.62	3.1159	0.71971	1.0241	330.14
280	1	12.502	323.86	403.85	3.1531	0.71975	1.0227	336.36
290	1	12.058	331.14	414.07	3.189	0.71989	1.0215	342.45
300	1	11.645	338.41	424.28	3.2236	0.72014	1.0206	348.41
400	1	8.6923	411.2	526.25	3.5169	0.72755	1.0213	402.28
500	1	6.9443	484.94	628.94	3.746	0.74317	1.0341	448.4
600	1	5.7849	560.44	733.31	3.9362	0.76446	1.054	489.11
700	1	4.9586	638.17	839.84	4.1004	0.78819	1.0769	525.9
800	1	4.3394	718.25	948.7	4.2457	0.81187	1.1001	559.79
900	1	3.858	800.62	1059.8	4.3766	0.83406	1.122	591.47
1000	1	3.4729	885.08	1173	4.4959	0.8541	1.1418	621.4
2.0 MPa								
60.106	2	959.35	−36.517	−34.432	−0.52806	1.1746	1.8945	1037.2
70	2	918.37	−17.862	−15.684	−0.23931	1.1055	1.8978	957.04
80	2	874.85	1.1029	3.389	0.015345	1.0482	1.9209	870.07
90	2	827.91	20.421	22.836	0.24434	1.0011	1.9752	775.98
100	2	775.21	40.496	43.076	0.45748	0.96337	2.0854	672.37
110	2	711.88	62.143	64.952	0.66581	0.93684	2.3236	554.28
118.52	2	638.95	83.537	86.667	0.8557	0.93119	2.8933	430.36
119.94	2	86.218	187.79	210.99	1.8984	0.93934	2.3646	180.47
120	2	86.053	187.89	211.14	1.8996	0.93796	2.3537	180.65
130	2	69.193	201.04	229.95	2.0505	0.82728	1.6066	202.61
140	2	59.893	211.37	244.76	2.1604	0.78807	1.3866	218.25
150	2	53.475	220.63	258.03	2.252	0.76725	1.2778	231.26
160	2	48.619	229.32	270.46	2.3322	0.7543	1.2125	242.71
170	2	44.747	237.66	282.35	2.4044	0.74559	1.1694	253.11
180	2	41.552	245.75	293.89	2.4703	0.73946	1.1389	262.73
190	2	38.851	253.68	305.16	2.5312	0.73501	1.1165	271.74
200	2	36.526	261.48	316.23	2.5881	0.73171	1.0994	280.26
210	2	34.496	269.18	327.16	2.6414	0.72921	1.0861	288.37
220	2	32.703	276.81	337.96	2.6916	0.72731	1.0755	296.14
230	2	31.105	284.38	348.67	2.7392	0.72586	1.0669	303.6
240	2	29.669	291.9	359.31	2.7845	0.72474	1.0599	310.81
250	2	28.369	299.38	369.88	2.8276	0.72391	1.0541	317.78
260	2	27.187	306.83	380.39	2.8689	0.72329	1.0493	324.54
270	2	26.105	314.25	390.86	2.9084	0.72287	1.0452	331.12
280	2	25.111	321.65	401.3	2.9464	0.72261	1.0418	337.52
290	2	24.194	329.04	411.7	2.9829	0.72251	1.039	343.76
300	2	23.345	336.41	422.08	3.018	0.72253	1.0366	349.86
400	2	17.346	409.87	525.17	3.3147	0.72881	1.0292	404.45
500	2	13.841	483.96	628.47	3.5451	0.74401	1.0388	450.76
600	2	11.527	559.7	733.21	3.736	0.76507	1.0571	491.48
700	2	9.8806	637.58	840	3.9006	0.78867	1.0791	528.22
800	2	8.6482	717.78	949.04	4.0462	0.81227	1.1017	562.05

(continued)

Table 8.6 (continued)

Temperature (K)	Pressure (MPa)	Density (kg/m^3)	Internal energy (kJ/kg)	Enthalpy (kJ/kg)	Entropy (kJ/kg K)	C_v (kJ/kg K)	C_p (kJ/kg K)	Sound speed (m/s)
900	2	7.6904	800.23	1060.3	4.1772	0.83439	1.1232	593.65
1000	2	6.9242	884.76	1173.6	4.2966	0.85438	1.1427	623.5
5.0 MPa								
60.637	5	961.65	−36.293	−31.094	−0.52449	1.1753	1.8852	1047.5
70	5	923.86	−18.87	−13.458	−0.25403	1.1108	1.8839	974.49
80	5	881.87	−0.23427	5.4355	−0.00177	1.0538	1.898	891.97
90	5	837.26	18.606	24.578	0.22366	1.0065	1.9352	804.31
100	5	788.5	37.918	44.259	0.43096	0.96786	2.0086	710.48
110	5	732.94	58.139	64.961	0.62816	0.93791	2.1463	608.94
120	5	665.22	80.13	87.646	0.82537	0.91918	2.4285	496.74
130	5	569.27	106.41	115.2	1.0455	0.92449	3.2807	363.04
140	5	321.4	156.88	172.44	1.4666	1.0484	8.5134	199.47
150	5	188.2	191.67	218.23	1.7847	0.88589	2.7271	214.84
160	5	151.41	207.61	240.63	1.9295	0.82675	1.9157	231.77
170	5	130.79	219.85	258.08	2.0354	0.79684	1.6127	245.87
180	5	116.74	230.51	273.34	2.1227	0.77833	1.4531	258.18
190	5	106.25	240.29	287.34	2.1984	0.76574	1.3546	269.25
200	5	97.957	249.5	300.54	2.2661	0.75671	1.288	279.38
210	5	91.163	258.32	313.17	2.3277	0.75002	1.2401	288.8
220	5	85.447	266.86	325.38	2.3846	0.74493	1.2043	297.62
230	5	80.542	275.2	337.28	2.4375	0.74101	1.1766	305.97
240	5	76.267	283.37	348.93	2.487	0.73794	1.1547	313.91
250	5	72.496	291.42	360.39	2.5338	0.73553	1.1371	321.51
260	5	69.135	299.36	371.68	2.5781	0.73364	1.1226	328.8
270	5	66.113	307.22	382.85	2.6203	0.73215	1.1106	335.83
280	5	63.378	315.01	393.9	2.6605	0.73101	1.1005	342.62
290	5	60.886	322.74	404.86	2.6989	0.73015	1.092	349.2
300	5	58.604	330.43	415.75	2.7358	0.72953	1.0848	355.59
400	5	43.02	405.92	522.15	3.0421	0.73248	1.0521	411.64
500	5	34.221	481.09	627.2	3.2766	0.74645	1.0522	458.24
600	5	28.484	557.5	733.03	3.4695	0.76688	1.0658	498.87
700	5	24.424	635.84	840.56	3.6352	0.7901	1.0852	535.39
800	5	21.39	716.37	950.13	3.7815	0.81344	1.1062	568.94
900	5	19.034	799.08	1061.8	3.9129	0.83538	1.1266	600.27
1000	5	17.148	883.81	1175.4	4.0326	0.85523	1.1454	629.86
10.0 MPa								
61.515	10	965.38	−35.9	−25.542	−0.51856	1.1767	1.871	1063.9
70	10	932.47	−20.424	−9.7	−0.2773	1.1195	1.8643	1001.5
80	10	892.64	−2.2552	8.9475	−0.0283	1.063	1.8673	925.08
90	10	851.12	15.948	27.697	0.19252	1.0158	1.8854	845.73
100	10	807.08	34.332	46.722	0.39293	0.97647	1.9235	763.38
110	10	759.34	53.089	66.259	0.57908	0.94426	1.9891	678.24
120	10	706.28	72.475	86.634	0.75631	0.91902	2.0939	591.06
130	10	645.54	92.842	108.33	0.92991	0.90116	2.2583	503.4

(continued)

Table 8.6 (continued)

Temperature (K)	Pressure (MPa)	Density (kg/m^3)	Internal energy (kJ/kg)	Enthalpy (kJ/kg)	Entropy (kJ/kg K)	C_v (kJ/kg K)	C_p (kJ/kg K)	Sound speed (m/s)
140	10	573.8	114.69	132.11	1.106	0.89141	2.5162	418.08
150	10	488.09	138.44	158.93	1.2909	0.88839	2.8417	343.38
160	10	397.09	162.79	187.98	1.4783	0.87851	2.8813	297.58
170	10	323.53	184.14	215.05	1.6426	0.85284	2.4971	282.66
180	10	273.4	201.35	237.92	1.7734	0.82651	2.0972	283.04
190	10	239.06	215.61	257.44	1.879	0.80601	1.825	288.78
200	10	214.14	228.03	274.73	1.9678	0.79063	1.6456	296.27
210	10	195.09	239.28	290.53	2.0449	0.77896	1.5224	304.27
220	10	179.92	249.71	305.29	2.1136	0.76993	1.4338	312.31
230	10	167.48	259.57	319.28	2.1758	0.76284	1.3677	320.23
240	10	157.02	269.01	332.7	2.2329	0.75721	1.3168	327.93
250	10	148.06	278.12	345.66	2.2858	0.75269	1.2767	335.4
260	10	140.28	286.97	358.26	2.3352	0.74904	1.2444	342.64
270	10	133.43	295.62	370.56	2.3816	0.74608	1.218	349.65
280	10	127.34	304.1	382.63	2.4255	0.74369	1.1961	356.45
290	10	121.87	312.44	394.5	2.4672	0.74176	1.1777	363.04
300	10	116.93	320.68	406.19	2.5068	0.74022	1.1622	369.45
400	10	84.586	399.59	517.82	2.8285	0.73825	1.0876	425.53
500	10	67.076	476.5	625.58	3.069	0.75032	1.0729	471.76
600	10	55.829	553.97	733.09	3.265	0.76976	1.0794	511.84
700	10	47.915	633.05	841.75	3.4324	0.79238	1.0947	547.77
800	10	42.012	714.11	952.14	3.5798	0.81532	1.1132	580.76
900	10	37.429	797.22	1064.4	3.712	0.83697	1.1319	611.56
1000	10	33.76	882.28	1178.5	3.8322	0.85661	1.1495	640.67
50.0 MPa								
68.213	50	991.73	−32.121	18.296	−0.47265	1.1913	1.7923	1172.6
70	50	986.23	−29.204	21.494	−0.42638	1.1807	1.7872	1163.2
80	50	955.77	−13.085	39.229	−0.18952	1.1273	1.7603	1111.4
90	50	925.77	2.6998	56.709	0.016395	1.082	1.7362	1061
100	50	896.16	18.169	73.963	0.1982	1.0432	1.715	1012.3
110	50	866.92	33.341	91.017	0.36076	1.0098	1.6962	965.44
120	50	838.02	48.228	107.89	0.50761	0.98087	1.6793	920.79
130	50	809.47	62.838	124.61	0.6414	0.95578	1.6637	878.66
140	50	781.31	77.173	141.17	0.76414	0.93393	1.6488	839.34
150	50	753.59	91.234	157.58	0.8774	0.91485	1.634	803.1
160	50	726.42	105.02	173.85	0.98237	0.89813	1.6187	770.11
170	50	699.88	118.51	189.96	1.08	0.88342	1.6027	740.43
180	50	674.07	131.72	205.9	1.1712	0.87041	1.5858	714.02
190	50	649.1	144.64	221.67	1.2564	0.85887	1.5678	690.78
200	50	625.05	157.26	237.25	1.3364	0.84858	1.5489	670.51
210	50	601.99	169.59	252.64	1.4115	0.83936	1.5292	652.99
220	50	579.96	181.62	267.83	1.4821	0.83109	1.5088	638
230	50	559	193.37	282.82	1.5487	0.82365	1.4879	625.28
240	50	539.12	204.85	297.59	1.6116	0.81695	1.4668	614.61

(continued)

Table 8.6 (continued)

Temperature (K)	Pressure (MPa)	Density (kg/m^3)	Internal energy (kJ/kg)	Enthalpy (kJ/kg)	Entropy (kJ/kg K)	C_v (kJ/kg K)	C_p (kJ/kg K)	Sound speed (m/s)
250	50	520.31	216.06	312.15	1.6711	0.81092	1.4457	605.74
260	50	502.56	227.02	326.51	1.7274	0.8055	1.4249	598.48
270	50	485.32	237.74	340.65	1.7808	0.80062	1.4046	592.6
280	50	470.06	248.23	354.6	1.8315	0.79625	1.385	587.94
290	50	455.23	258.52	368.36	1.8798	0.79233	1.3662	584.31
300	50	441.26	268.62	381.93	1.9258	0.78885	1.3483	581.59
400	50	338.31	361.93	509.72	2.2943	0.77153	1.2228	582.79
500	50	276.19	447.79	628.82	2.5603	0.77471	1.1674	603.99
600	50	234.52	531.27	744.39	2.771	0.78877	1.148	629.42
700	50	204.53	614.67	859.01	2.9477	0.80783	1.1464	655.37
800	50	181.34	698.99	973.95	3.1012	0.82827	1.1533	680.9
900	50	163.37	784.65	1089.8	3.2376	0.84808	1.1636	705.77
1000	50	149.27	871.75	1206.7	3.3608	0.8663	1.1751	729.97
100.0 MPa								
75.918	100	1019.1	−26.424	71.703	−0.4214	1.2066	1.7381	1281.7
80	100	1008.9	−20.352	78.77	−0.33071	1.1866	1.7249	1265.8
90	100	984.31	−5.7315	95.862	−0.12936	1.1423	1.6938	1228.1
100	100	960.52	8.5432	112.65	0.047572	1.1041	1.6647	1192.2
110	100	937.45	22.49	129.16	0.20494	1.0708	1.6375	1158.2
120	100	915.07	36.128	145.41	0.34632	1.0416	1.6121	1125.9
130	100	893.35	49.472	161.41	0.47441	1.0159	1.5884	1095.4
140	100	872.27	62.538	177.18	0.5913	0.99325	1.5662	1066.8
150	100	851.81	75.34	192.74	0.69863	0.9731	1.5453	1039.9
160	100	831.95	87.891	208.09	0.79773	0.95513	1.5255	1014.9
170	100	812.68	100.2	223.25	0.88964	0.93906	1.5068	991.62
180	100	794	112.29	238.23	0.97526	0.92462	1.489	970.07
190	100	775.9	124.15	253.03	1.0553	0.91163	1.4719	950.19
200	100	758.36	135.81	267.67	1.1304	0.89989	1.4556	931.9
210	100	741.39	147.27	282.15	1.201	0.88927	1.44	915.14
220	100	724.97	158.54	296.47	1.2677	0.87963	1.4249	899.81
230	100	709.1	169.62	310.65	1.3307	0.87087	1.4105	885.83
240	100	693.76	180.54	324.68	1.3904	0.8629	1.3966	873.1
250	100	678.94	191.3	338.58	1.4472	0.85564	1.3833	861.54
260	100	664.64	201.89	352.35	1.5012	0.84902	1.3705	851.07
270	100	650.83	212.35	366	1.5527	0.84299	1.3583	841.59
280	100	637.51	222.66	379.52	1.6018	0.83748	1.3465	833.03
290	100	624.66	232.84	392.93	1.6489	0.83247	1.3353	825.32
300	100	612.26	242.9	406.23	1.694	0.82791	1.3245	818.38
400	100	509.9	338.08	534.2	2.0627	0.80139	1.2421	779.71
500	100	437.06	427.11	655.91	2.3345	0.79808	1.1975	772.32
600	100	383.23	513.6	774.54	2.5508	0.80767	1.1784	778.15
700	100	341.89	599.59	892.08	2.732	0.82358	1.1742	790.05
800	100	309.08	686.1	1009.6	2.889	0.8417	1.1779	805.03
900	100	282.36	773.62	1127.8	3.0281	0.85974	1.1853	821.68
1000	100	260.12	862.3	1246.7	3.1535	0.87659	1.194	839.25

Reference

1. E.W. Lemmon, R.T. Jacobsen, S.G. Penoncello, D.G. Friend, Thermodynamic properties of air and mixtures of nitrogen, argon, and oxygen from 60 to 2000 K at pressures to 2000 MPa. J. Phys. Chem. Ref. Data. **29**(3), 331–385 (2000)

Chapter 9
Carbon Monoxide

9.1 Formulation for Carbon Monoxide

The formulation for carbon monoxide was developed by Lemmon and Span [1]. The uncertainty statement in REFPROP is as follows.

The equation of state is valid from the triple point to 500 K with pressures to 100 MPa. At higher pressures, the deviations from the equation increase rapidly and it is not recommended to use the equation above 100 MPa. The uncertainties in the equation are 0.3% in density (approaching 1% near the critical point), 0.2% in vapor pressure, and 2% in heat capacities. The uncertainty in the speed of sound is unknown (Tables 9.1, 9.2, 9.3 and 9.4).

J.W. Leachman et al., *Thermodynamic Properties of Cryogenic Fluids*,
International Cryogenics Monograph Series, DOI 10.1007/978-3-319-57835-4_9

Table 9.1 Fixed point properties and correlation limits for carbon monoxide

	Temperature (K)	Pressure (kPa)	Density (mol/dm^3)
Critical point	132.86	3494.0	10.85
Triple point	68.16	15.45	30.33
Upper limit	500	100,000	33.84
Lower limit	68.16		

Table 9.2 Fluid constants and normal boiling point for carbon monoxide

Molar mass	28.0101 g/mol
Ideal-gas constant	8.314472 J/(mol K)
Pitzer's acentric factor	0.0497
Normal boiling point	81.64 K

Table 9.3 Fundamental equation of state coefficients and exponents for carbon monoxide

i	N_i	t_i	d_i	p_i
1	0.90554	0.25	1	
2	−2.4515	1.125	1	
3	0.53149	1.5	1	
4	0.024173	1.375	2	
5	0.072156	0.25	3	
6	0.00018818	0.875	7	
7	0.19405	0.625	2	1
8	−0.043268	1.75	5	1
9	−0.12778	3.625	1	2
10	−0.027896	3.625	4	2
11	−0.034154	14.5	3	3
12	0.016329	12	4	3

Table 9.4 Ancillary function coefficients and exponents for carbon monoxide

	α^0	
i	a_i	k_i
0	3.5	
1	−3.3728318564	
2	3.3683460039	
3	2.2311×10^{-7}	1.5
4	1.0128	3089

$m = 3$, $n = 4$

	Vapor pressure		Saturated liquid density		Saturated vapor density	
i	N_i	k_i	N_i	k_i	N_i	k_i
1	−6.1192	1	2.957	0.398	−2.5439	0.395
2	1.0411	1.5	−4.288	0.735	−5.5601	1.21
3	−6.2162	3.9	8.7643	1.08	−8.5276	3
4	10.437	4.5	−8.4001	1.5	−5.1163	3.5
5	−7.6813	5.4	3.6372	1.9	−17.701	6
6					−29.858	8

9.2 Property Plots for Carbon Monoxide

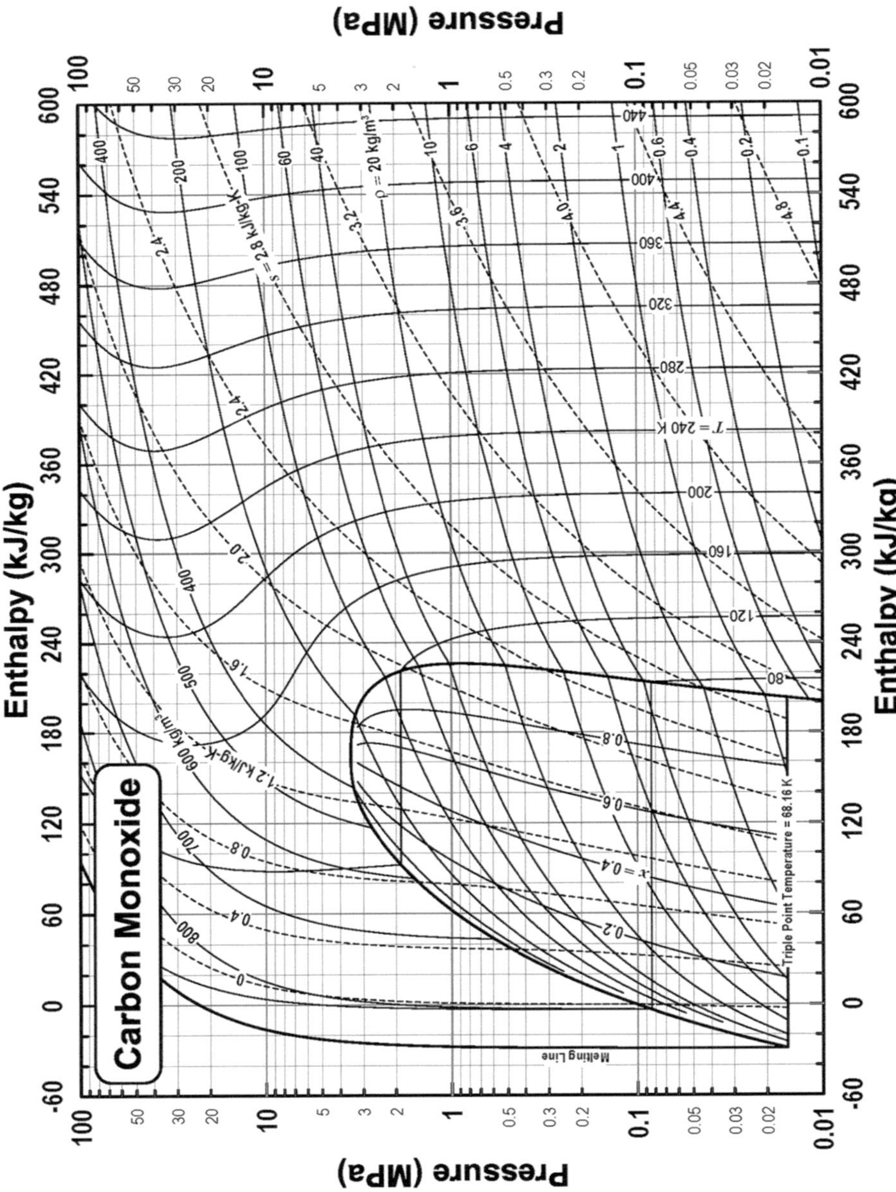
Carbon Monoxide
Enthalpy (kJ/kg)
Pressure (MPa)
Melting Line
Triple Point Temperature = 68.16 K
ρ = 20 kg/m³
s = 2.8 kJ/kg-K
T = 240 K
x = 0.4

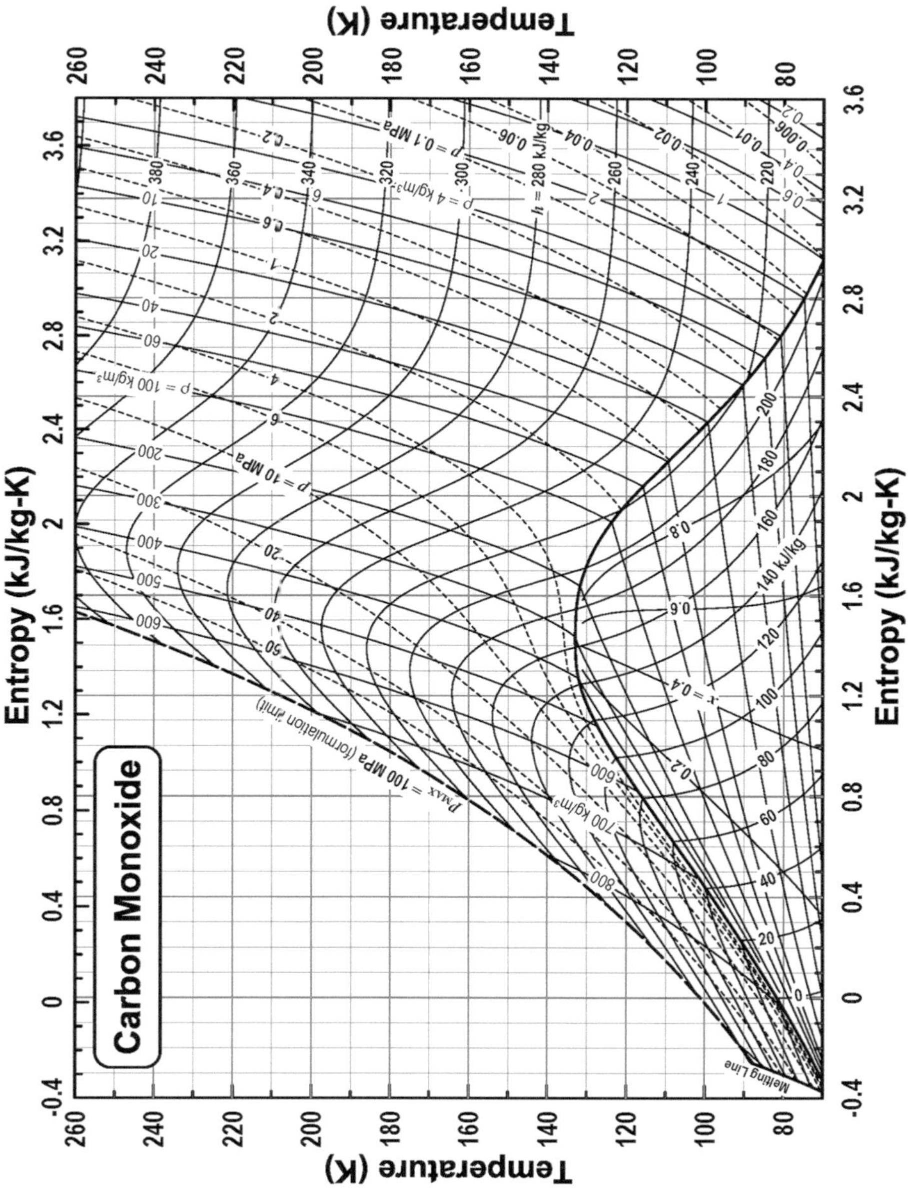
Carbon Monoxide
Entropy (kJ/kg-K)
Temperature (K)
p = 0.1 MPa
p = 10 MPa
p_{MAX} = 100 MPa (formulation limit)
ρ = 4 kg/m³
ρ = 100 kg/m³
700 kg/m³
h = 280 kJ/kg
140 kJ/kg
x = 0.4
Melting Line

9.3 Thermodynamic Data Tables for Carbon Monoxide

See Tables 9.5 and 9.6.

Table 9.5 Thermodynamic properties of saturated carbon monoxide

Temperature (K)	Pressure (MPa)	Liquid density (kg/m^3)	Vapor density (kg/m^3)	Liquid enthalpy (kJ/kg)	Vapor enthalpy (kJ/kg)	Liquid entropy (kJ/kg K)	Vapor entropy (kJ/kg K)	Liquid C_v (kJ/kg K)	Vapor C_v (kJ/kg K)	Liquid sound speed (m/s)	Vapor sound speed (m/s)
68.16	0.015537	849.54	0.77609	−28.956	202.99	−0.38629	3.0167	1.2621	0.7529	998.2	167.25
70	0.021053	842.1	1.0267	−24.989	204.72	−0.32895	2.9527	1.2426	0.75527	980.5	169.22
73	0.033303	829.82	1.5658	−18.542	207.47	−0.23898	2.857	1.2132	0.75988	951.56	172.26
76	0.050599	817.34	2.3005	−12.108	210.1	−0.15288	2.7709	1.1864	0.76549	922.49	175.07
79	0.07422	804.62	3.2731	−5.6724	212.6	−0.07021	2.6928	1.1619	0.7722	893.24	177.62
81.638	0.10132	793.21	4.3612	0	214.68	0	2.6297	1.142	0.77907	867.32	179.65
82	0.10556	791.63	4.5294	0.77952	214.96	0.009462	2.6214	1.1394	0.78009	863.76	179.92
85	0.14612	778.33	6.1192	7.2624	217.15	0.086489	2.5558	1.1187	0.78921	833.99	181.93
88	0.19748	764.68	8.0966	13.791	219.16	0.1612	2.495	1.0995	0.79961	803.89	183.66
91	0.26127	750.62	10.522	20.382	220.97	0.23391	2.4382	1.0818	0.8113	773.39	185.1
94	0.33919	736.1	13.461	27.052	222.56	0.30489	2.3847	1.0655	0.8243	742.41	186.22
97	0.43296	721.06	16.992	33.82	223.89	0.37441	2.3339	1.0506	0.83864	710.87	187.04
100	0.54438	705.41	21.204	40.708	224.95	0.44275	2.2851	1.0369	0.85438	678.68	187.52
103	0.67523	689.06	26.203	47.739	225.69	0.51018	2.2379	1.0246	0.87163	645.69	187.68
106	0.82736	671.88	32.123	54.945	226.08	0.577	2.1915	1.0137	0.89057	611.77	187.51
109	1.0027	653.74	39.129	62.36	226.08	0.64352	2.1455	1.0046	0.9115	576.71	186.98
112	1.2031	634.41	47.444	70.032	225.6	0.71013	2.0991	0.99754	0.93488	540.25	186.11
115	1.4307	613.63	57.368	78.022	224.56	0.7773	2.0516	0.99319	0.96144	502.06	184.88
118	1.6877	590.99	69.337	86.416	222.83	0.84569	2.0018	0.9926	0.9923	461.63	183.27
121	1.9765	565.88	84.029	95.345	220.21	0.91623	1.9482	0.99758	1.0293	418.28	181.28
124	2.2997	537.2	102.6	105.03	216.36	0.99051	1.8883	1.0115	1.0758	370.88	178.84
127	2.6609	502.75	127.37	115.92	210.6	1.0717	1.8172	1.0419	1.1384	317.52	175.86
130	3.0647	456.24	164.79	129.27	201.08	1.169	1.7214	1.1102	1.2345	254.03	171.86
132.86	3.4982	303.91	303.91	164.71	164.71	1.4294	1.4294				

Table 9.6 Thermodynamic properties of carbon monoxide

Temperature (K)	Pressure (MPa)	Density (kg/m^3)	Internal energy (kJ/kg)	Enthalpy (kJ/kg)	Entropy (kJ/kg K)	C_v (kJ/kg K)	C_p (kJ/kg K)	Sound speed (m/s)
0.1 MPa								
70	0.1	842.24	−25.046	−24.927	−0.3294	1.2428	2.1497	981.06
80	0.1	800.36	−3.6377	−3.5128	−0.04345	1.1542	2.142	883.58
81.523	0.1	793.72	−0.37468	−0.24869	−0.00303	1.1429	2.1455	868.46
81.523	0.1	4.3085	191.38	214.59	2.6324	0.77875	1.1308	179.57
90	0.1	3.8593	198.1	224.01	2.7423	0.76214	1.0958	190
100	0.1	3.4446	205.83	234.86	2.8566	0.75395	1.0765	201.29
110	0.1	3.1145	213.46	245.57	2.9587	0.74977	1.0657	211.82
120	0.1	2.8441	221.03	256.19	3.0511	0.74734	1.059	221.75
130	0.1	2.6182	228.56	266.76	3.1357	0.74585	1.0546	231.19
140	0.1	2.4262	236.07	277.29	3.2137	0.74488	1.0515	240.21
150	0.1	2.261	243.56	287.79	3.2862	0.74424	1.0492	248.87
160	0.1	2.1171	251.04	298.27	3.3538	0.7438	1.0476	257.22
170	0.1	1.9907	258.51	308.74	3.4173	0.74349	1.0463	265.28
180	0.1	1.8786	265.97	319.2	3.4771	0.74328	1.0453	273.09
190	0.1	1.7736	273.42	329.65	3.5336	0.74312	1.0445	280.67
200	0.1	1.6838	280.88	340.09	3.5871	0.74301	1.0439	288.05
210	0.1	1.6077	288.32	350.52	3.638	0.74293	1.0433	295.23
220	0.1	1.5341	295.77	360.96	3.6866	0.74289	1.0429	302.23
230	0.1	1.4659	303.21	371.38	3.7329	0.74287	1.0426	309.07
240	0.1	1.4055	310.66	381.81	3.7773	0.74289	1.0423	315.76
250	0.1	1.349	318.1	392.23	3.8198	0.74293	1.0421	322.3
260	0.1	1.2968	325.54	402.65	3.8607	0.74301	1.042	328.71
270	0.1	1.2486	332.98	413.07	3.9	0.74312	1.0419	334.99
280	0.1	1.2038	340.42	423.49	3.9379	0.74328	1.0419	341.15
290	0.1	1.1622	347.86	433.91	3.9745	0.74349	1.042	347.19
300	0.1	1.1233	355.31	444.33	4.0098	0.74376	1.0422	353.12
400	0.1	0.84203	430.02	548.78	4.3102	0.75071	1.0483	407.29
500	0.1	0.67351	505.85	654.33	4.5457	0.76684	1.0641	454
0.2 MPa								
68.18	0.2	849.77	−29.001	−28.766	−0.38668	1.2623	2.1564	999.26
70	0.2	842.42	−25.086	−24.848	−0.32998	1.243	2.1491	981.76
80	0.2	800.6	−3.6918	−3.442	−0.04412	1.1544	2.141	884.44
88.131	0.2	764.07	13.817	14.078	0.16443	1.0987	2.175	802.56
88.131	0.2	8.1929	194.84	219.25	2.4924	0.80009	1.1933	183.73
90	0.2	7.9849	196.41	221.46	2.5173	0.79127	1.1747	186.24
100	0.2	7.0549	204.54	232.89	2.6378	0.76747	1.1203	198.6
110	0.2	6.3394	212.41	243.96	2.7432	0.75784	1.0952	209.78
120	0.2	5.7653	220.14	254.83	2.8379	0.75269	1.0805	220.16
130	0.2	5.2919	227.79	265.58	2.9239	0.74959	1.071	229.93
140	0.2	4.8934	235.39	276.26	3.0031	0.74761	1.0644	239.21
150	0.2	4.5527	242.95	286.88	3.0763	0.7463	1.0598	248.07
160	0.2	4.2577	250.49	297.46	3.1446	0.74539	1.0563	256.58
170	0.2	3.9995	258	308.01	3.2086	0.74476	1.0536	264.78

(continued)

Table 9.6 (continued)

Temperature (K)	Pressure (MPa)	Density (kg/m^3)	Internal energy (kJ/kg)	Enthalpy (kJ/kg)	Entropy (kJ/kg K)	C_v (kJ/kg K)	C_p (kJ/kg K)	Sound speed (m/s)
180	0.2	3.7714	265.5	318.53	3.2687	0.7443	1.0516	272.7
190	0.2	3.5684	272.99	329.04	3.3255	0.74397	1.05	280.37
200	0.2	3.3865	280.48	339.53	3.3794	0.74372	1.0486	287.82
210	0.2	3.2224	287.95	350.01	3.4305	0.74355	1.0476	295.06
220	0.2	3.0738	295.42	360.49	3.4792	0.74342	1.0467	302.12
230	0.2	2.9383	302.88	370.95	3.5257	0.74334	1.046	309.01
240	0.2	2.8145	310.34	381.41	3.5702	0.7433	1.0454	315.74
250	0.2	2.7007	317.8	391.86	3.6129	0.7433	1.0449	322.31
260	0.2	2.5959	325.26	402.3	3.6539	0.74334	1.0445	328.75
270	0.2	2.4989	332.71	412.75	3.6933	0.74343	1.0442	335.06
280	0.2	2.409	340.17	423.19	3.7312	0.74357	1.0441	341.24
290	0.2	2.3254	347.62	433.63	3.7679	0.74376	1.044	347.3
300	0.2	2.2474	355.08	444.07	3.8033	0.744	1.044	353.25
400	0.2	1.6837	429.86	548.65	4.1041	0.75086	1.0493	407.52
500	0.2	1.3465	505.74	654.27	4.3397	0.76694	1.0647	454.26
0.5 MPa								
68.248	0.5	850.01	−28.969	−28.381	−0.38621	1.2621	2.1545	1000.6
70	0.5	842.95	−25.205	−24.612	−0.33169	1.2436	2.1474	983.86
80	0.5	801.31	−3.8529	−3.229	−0.04614	1.1551	2.1379	887.02
90	0.5	756.2	17.666	18.327	0.2077	1.0881	2.1833	786.48
98.864	0.5	711.41	37.381	38.084	0.417	1.0419	2.2877	690.95
98.864	0.5	19.522	198.97	224.58	2.3034	0.84825	1.3637	187.38
100	0.5	19.189	200.06	226.11	2.3188	0.83708	1.3367	189.18
110	0.5	16.806	208.98	238.73	2.4391	0.78802	1.2094	203.14
120	0.5	15.053	217.31	250.53	2.5419	0.77046	1.1564	215.18
130	0.5	13.678	225.38	261.93	2.6331	0.7615	1.1264	226.08
140	0.5	12.558	233.28	273.09	2.7159	0.75612	1.1071	236.19
150	0.5	11.623	241.07	284.09	2.7917	0.75264	1.0937	245.69
160	0.5	10.826	248.79	294.98	2.862	0.75028	1.0841	254.7
170	0.5	10.138	256.46	305.78	2.9275	0.74862	1.0768	263.3
180	0.5	9.5366	264.09	316.52	2.9889	0.74742	1.0713	271.54
190	0.5	9.0056	271.69	327.21	3.0467	0.74655	1.0669	279.49
200	0.5	8.5328	279.26	337.86	3.1013	0.74589	1.0634	287.16
210	0.5	8.1088	286.82	348.48	3.1531	0.7454	1.0605	294.6
220	0.5	7.7262	294.36	359.07	3.2024	0.74503	1.0582	301.82
230	0.5	7.3791	301.89	369.65	3.2494	0.74475	1.0562	308.85
240	0.5	7.0625	309.4	380.2	3.2943	0.74456	1.0546	315.69
250	0.5	6.7726	316.91	390.74	3.3373	0.74443	1.0533	322.38
260	0.5	6.506	324.41	401.27	3.3786	0.74436	1.0521	328.9
270	0.5	6.26	331.91	411.78	3.4183	0.74436	1.0512	335.29
280	0.5	6.0323	339.4	422.29	3.4565	0.74442	1.0504	341.54
290	0.5	5.8208	346.89	432.79	3.4934	0.74455	1.0498	347.66
300	0.5	5.6238	354.38	443.29	3.529	0.74474	1.0494	353.67
400	0.5	4.2064	429.38	548.25	3.8309	0.75129	1.0521	408.22

(continued)

Table 9.6 (continued)

Temperature (K)	Pressure (MPa)	Density (kg/m³)	Internal energy (kJ/kg)	Enthalpy (kJ/kg)	Entropy (kJ/kg K)	C_v (kJ/kg K)	C_p (kJ/kg K)	Sound speed (m/s)
500	0.5	3.3624	505.39	654.09	4.067	0.76725	1.0664	455.06
1.0 MPa								
68.361	1	850.41	−28.915	−27.739	−0.38543	1.262	2.1514	1002.9
70	1	843.84	−25.403	−24.218	−0.33452	1.2447	2.1445	987.33
80	1	802.48	−4.1185	−2.8723	−0.04948	1.1561	2.133	891.28
90	1	757.81	17.297	18.616	0.20358	1.0891	2.1743	791.9
100	1	707.58	39.44	40.853	0.43776	1.0376	2.2889	685.44
110	1	38.266	201.66	227.8	2.1618	0.8934	1.6196	189.11
108.96	1	654	60.724	62.253	0.64257	1.0047	2.5215	577.22
108.96	1	39.021	200.45	226.08	2.1462	0.91118	1.6767	186.99
120	1	32.814	211.89	242.37	2.2888	0.81173	1.3489	205.87
130	1	29.12	220.95	255.29	2.3923	0.78488	1.248	219.29
140	1	26.336	229.5	267.47	2.4826	0.77173	1.1938	231.06
150	1	24.121	237.77	279.23	2.5637	0.76391	1.1596	241.75
160	1	22.299	245.86	290.7	2.6378	0.75881	1.1362	251.65
170	1	20.763	253.81	301.98	2.7061	0.75529	1.1193	260.95
180	1	19.445	261.68	313.1	2.7697	0.75279	1.1067	269.75
190	1	18.299	269.47	324.12	2.8293	0.75096	1.0969	278.15
200	1	17.29	277.21	335.05	2.8853	0.74959	1.0892	286.2
210	1	16.393	284.91	345.91	2.9383	0.74855	1.0831	293.95
220	1	15.591	292.57	356.71	2.9886	0.74775	1.078	301.43
230	1	14.867	300.21	367.47	3.0364	0.74714	1.0738	308.68
240	1	14.211	307.82	378.19	3.082	0.74667	1.0704	315.72
250	1	13.613	315.42	388.88	3.1257	0.74633	1.0674	322.57
260	1	13.065	323	399.54	3.1675	0.74608	1.065	329.24
270	1	12.561	330.57	410.18	3.2076	0.74593	1.0629	335.75
280	1	12.096	338.13	420.8	3.2463	0.74586	1.0611	342.11
290	1	11.665	345.67	431.4	3.2835	0.74587	1.0596	348.33
300	1	11.264	353.22	441.99	3.3194	0.74597	1.0584	354.42
400	1	8.4028	428.6	547.61	3.6232	0.75201	1.0567	409.43
500	1	6.7119	504.81	653.8	3.8601	0.76776	1.0692	456.39
2.0 MPa								
68.586	2	851.2	−28.806	−26.456	−0.38386	1.2616	2.1453	1007.5
70	2	845.59	−25.792	−23.426	−0.34013	1.2467	2.1391	994.19
80	2	804.77	−4.6384	−2.1533	−0.05604	1.1582	2.1237	899.63
90	2	760.94	16.581	19.209	0.19553	1.0912	2.1575	802.44
100	2	712.14	38.397	41.205	0.42719	1.0392	2.2551	699.62
110	2	654.53	61.651	64.707	0.65102	1.0013	2.4765	585.04
120	2	576.65	88.594	92.062	0.88864	0.993	3.1524	438.55
121.23	2	563.82	92.511	96.058	0.92177	0.99827	3.3539	414.8
121.23	2	85.297	196.52	219.97	1.9439	1.0325	2.8155	181.11
130	2	68.897	209.61	238.64	2.0929	0.86364	1.7589	203.43
140	2	59.041	220.66	254.53	2.2109	0.81189	1.469	220.35
150	2	52.451	230.39	268.52	2.3074	0.78989	1.3418	234.04

(continued)

Table 9.6 (continued)

Temperature (K)	Pressure (MPa)	Density (kg/m^3)	Internal energy (kJ/kg)	Enthalpy (kJ/kg)	Entropy (kJ/kg K)	C_v (kJ/kg K)	C_p (kJ/kg K)	Sound speed (m/s)
160	2	47.536	239.47	281.54	2.3915	0.77751	1.269	245.98
170	2	43.652	248.17	293.98	2.4669	0.76955	1.2217	256.76
180	2	40.467	256.6	306.03	2.5358	0.76407	1.1887	266.7
190	2	37.788	264.86	317.79	2.5994	0.76014	1.1645	276
200	2	35.49	272.98	329.34	2.6586	0.75723	1.1461	284.77
210	2	33.489	281	340.72	2.7142	0.75502	1.1317	293.12
220	2	31.727	288.94	351.98	2.7665	0.75333	1.1201	301.1
230	2	30.159	296.81	363.13	2.8161	0.75201	1.1108	308.76
240	2	28.752	304.64	374.2	2.8632	0.75098	1.1031	316.15
250	2	27.481	312.42	385.2	2.9081	0.75018	1.0966	323.3
260	2	26.326	320.17	396.14	2.951	0.74955	1.0912	330.23
270	2	25.271	327.88	407.02	2.9921	0.74909	1.0866	336.96
280	2	24.302	335.57	417.87	3.0316	0.74875	1.0827	343.52
290	2	23.409	343.24	428.68	3.0695	0.74854	1.0794	349.91
300	2	22.583	350.9	439.46	3.106	0.74845	1.0765	356.15
400	2	16.762	427.04	546.36	3.4136	0.75345	1.0658	411.94
500	2	13.371	503.68	653.26	3.6521	0.76877	1.0746	459.1
5.0 MPa								
69.255	5	853.53	−28.473	−22.615	−0.37919	1.2607	2.1279	1020.8
70	5	850.67	−26.91	−21.032	−0.35645	1.2529	2.1242	1014.1
80	5	811.33	−6.1166	0.04607	−0.07496	1.1644	2.0987	923.59
90	5	769.7	14.582	21.078	0.17274	1.0973	2.1145	832.06
100	5	724.47	35.582	42.484	0.39822	1.0447	2.1751	737.92
110	5	673.55	57.371	64.794	0.61077	1.0036	2.3011	638.82
120	5	612.89	80.768	88.926	0.82058	0.97448	2.558	530.54
130	5	531.61	107.83	117.23	1.0468	0.96946	3.2366	402.59
140	5	363.44	150.73	164.49	1.395	1.0835	7.9055	229.21
150	5	193.24	196.25	222.13	1.7952	0.92378	3.2714	220.33
160	5	151.18	214.61	247.69	1.9605	0.84896	2.1099	237.25
170	5	129.21	227.94	266.64	2.0755	0.81813	1.7322	251.39
180	5	114.67	239.33	282.93	2.1687	0.80067	1.5444	263.79
190	5	103.97	249.68	297.77	2.249	0.7892	1.4317	274.99
200	5	95.599	259.39	311.69	2.3204	0.78107	1.3567	285.27
210	5	88.792	268.67	324.98	2.3852	0.77507	1.3033	294.83
220	5	83.097	277.63	337.8	2.4449	0.77049	1.2636	303.82
230	5	78.23	286.37	350.28	2.5003	0.76694	1.2329	312.32
240	5	74.004	294.92	362.48	2.5523	0.76413	1.2087	320.42
250	5	70.286	303.33	374.47	2.6012	0.7619	1.1891	328.16
260	5	66.981	311.63	386.28	2.6475	0.76011	1.1731	335.61
270	5	64.016	319.83	397.94	2.6915	0.75867	1.1597	342.78
280	5	61.336	327.96	409.48	2.7335	0.75752	1.1485	349.71
290	5	58.899	336.02	420.91	2.7736	0.75661	1.1389	356.43
300	5	56.67	344.03	432.26	2.8121	0.75593	1.1308	362.95
400	5	41.522	422.49	542.91	3.1307	0.75774	1.0919	420.15

(continued)

Table 9.6 (continued)

Temperature (K)	Pressure (MPa)	Density (kg/m^3)	Internal energy (kJ/kg)	Enthalpy (kJ/kg)	Entropy (kJ/kg K)	C_v (kJ/kg K)	C_p (kJ/kg K)	Sound speed (m/s)
500	5	33.013	500.37	651.83	3.3737	0.77179	1.0901	467.56
10.0 MPa								
70.356	10	857.28	−27.899	−16.235	−0.37152	1.2592	2.1013	1042.3
80	10	821.36	−8.347	3.8279	−0.10424	1.1744	2.0653	960.41
90	10	782.67	11.655	24.432	0.13845	1.1074	2.0606	876.19
100	10	741.78	31.656	45.137	0.35658	1.0546	2.0853	792.04
110	10	697.76	51.909	66.241	0.55767	1.0121	2.1409	707.52
120	10	649.4	72.683	88.082	0.74765	0.97769	2.2346	622.44
130	10	595.01	94.299	111.11	0.93186	0.95045	2.3804	537.3
140	10	532.35	117.14	135.92	1.1157	0.92982	2.5961	454.21
150	10	459.72	141.45	163.21	1.3038	0.91377	2.8575	379.46
160	10	381.89	166.27	192.45	1.4925	0.89518	2.9279	326.64
170	10	314.75	188.61	220.38	1.6619	0.87017	2.612	303.35
180	10	266.49	206.96	244.49	1.7999	0.84682	2.2228	298.7
190	10	232.72	222.23	265.2	1.9119	0.82878	1.9371	301.79
200	10	208.11	235.49	283.54	2.006	0.81513	1.7433	307.84
210	10	189.3	247.43	300.26	2.0876	0.80463	1.609	315.02
220	10	174.36	258.49	315.84	2.1601	0.79641	1.5122	322.57
230	10	162.13	268.91	330.59	2.2257	0.78988	1.4399	330.17
240	10	151.88	278.85	344.69	2.2857	0.78462	1.3843	337.68
250	10	143.11	288.44	358.31	2.3413	0.78034	1.3405	345.03
260	10	135.51	297.74	371.53	2.3932	0.77683	1.3052	352.2
270	10	128.83	306.81	384.44	2.4419	0.77394	1.2763	359.17
280	10	122.89	315.7	397.08	2.4879	0.77155	1.2524	365.96
290	10	117.58	324.44	409.5	2.5314	0.76959	1.2322	372.57
300	10	112.78	333.06	421.73	2.5729	0.76797	1.2151	379.01
400	10	81.45	415.34	538.11	2.9083	0.76473	1.1313	435.68
500	10	64.563	495.18	650.06	3.1582	0.7767	1.1135	482.48
50.0 MPa								
78.556	50	882.92	−22.813	33.817	−0.31635	1.2522	1.9599	1188.9
80	50	878.9	−20.25	36.639	−0.28074	1.2409	1.9507	1180.2
90	50	851.32	−2.8635	55.869	−0.05419	1.1736	1.8982	1121.8
100	50	824.11	13.978	74.65	0.14372	1.1209	1.8598	1066.8
110	50	797.21	30.376	93.095	0.31954	1.0784	1.8303	1015.1
120	50	770.62	46.39	111.27	0.47773	1.0433	1.806	966.87
130	50	744.35	62.052	129.22	0.62143	1.0139	1.7846	922.05
140	50	718.48	77.377	146.97	0.75293	0.98877	1.7642	880.75
150	50	693.09	92.368	164.51	0.87395	0.96718	1.7438	843.01
160	50	668.28	107.02	181.84	0.98582	0.94846	1.7225	808.85
170	50	644.15	121.33	198.95	1.0896	0.93211	1.7	778.2
180	50	620.8	135.29	215.84	1.1861	0.91775	1.6763	750.97
190	50	598.31	148.91	232.48	1.276	0.90506	1.6515	726.97
200	50	576.75	162.17	248.86	1.3601	0.89381	1.6259	706.01
210	50	556.16	175.09	264.99	1.4388	0.88378	1.5999	687.83

(continued)

Table 9.6 (continued)

Temperature (K)	Pressure (MPa)	Density (kg/m³)	Internal energy (kJ/kg)	Enthalpy (kJ/kg)	Entropy (kJ/kg K)	C_v (kJ/kg K)	C_p (kJ/kg K)	Sound speed (m/s)
220	50	536.57	187.68	280.86	1.5126	0.87481	1.574	672.18
230	50	517.98	199.94	296.47	1.582	0.86676	1.5483	658.8
240	50	500.38	211.91	311.83	1.6474	0.85952	1.5234	647.43
250	50	483.75	223.58	326.94	1.7091	0.853	1.4992	637.84
260	50	468.04	234.99	341.82	1.7674	0.8471	1.4761	629.81
270	50	453.21	246.15	356.47	1.8227	0.84178	1.4542	623.15
280	50	439.23	257.07	370.91	1.8752	0.83696	1.4333	617.68
290	50	426.04	267.78	385.14	1.9252	0.83262	1.4137	613.25
300	50	413.59	278.29	399.18	1.9728	0.8287	1.3953	609.73
400	50	320.42	375.52	531.57	2.3545	0.80756	1.2681	604.51
500	50	262.92	464.96	655.13	2.6305	0.80895	1.2116	622.3
100.0 MPa								
87.633	100	908.36	−15.911	94.177	−0.26142	1.2483	1.8671	1336.7
90	100	903.16	−12.143	98.58	−0.21184	1.2335	1.8534	1326
100	100	881.58	3.4251	116.86	−0.01923	1.1795	1.804	1282.4
110	100	860.64	18.5	134.69	0.15078	1.1359	1.7642	1241.8
120	100	840.29	33.156	152.16	0.30282	1.0999	1.7309	1204
130	100	820.49	47.444	169.32	0.4402	1.0696	1.7019	1168.8
140	100	801.23	61.402	186.21	0.56536	1.0438	1.6759	1136.2
150	100	782.51	75.055	202.85	0.68016	1.0214	1.652	1106
160	100	764.33	88.422	219.26	0.78605	1.0019	1.6295	1078.1
170	100	746.7	101.52	235.44	0.88419	0.98479	1.608	1052.5
180	100	729.62	114.36	251.42	0.97551	0.96958	1.5873	1029
190	100	713.09	126.96	267.19	1.0608	0.95601	1.5672	1007.4
200	100	697.1	139.32	282.77	1.1407	0.94384	1.5477	987.69
210	100	681.67	151.45	298.15	1.2157	0.93289	1.5287	969.71
220	100	666.79	163.37	313.34	1.2864	0.92298	1.5103	953.34
230	100	652.43	175.08	328.36	1.3532	0.91399	1.4924	938.46
240	100	638.61	186.6	343.19	1.4163	0.90582	1.4751	924.94
250	100	625.29	197.93	357.86	1.4762	0.89836	1.4585	912.67
260	100	612.47	209.09	372.36	1.5331	0.89155	1.4425	901.54
270	100	600.13	220.08	386.71	1.5872	0.88532	1.4271	891.46
280	100	588.25	230.91	400.91	1.6389	0.87962	1.4124	882.33
290	100	576.81	241.6	414.96	1.6882	0.8744	1.3985	874.07
300	100	565.8	252.14	428.88	1.7354	0.86963	1.3852	866.6
400	100	475.33	351.61	561.99	2.119	0.84096	1.2871	822.33
500	100	410.67	444.43	687.93	2.4002	0.83662	1.2382	808.91

Reference

1. E.W. Lemmon, R. Span, Short fundamental equations of state for 20 industrial fluids. J. Chem. Eng. Data **51**, 785–850 (2006)

Chapter 10
Fluorine

10.1 Formulation for Fluorine

The formulation for fluorine is one of the oldest within this text and published by de Reuck [1]. Although the equation was fitted with temperatures on the IPTS-68 temperature scale, the uncertainties in calculated properties are generally much higher than the change caused by use of the older temperature scale. The uncertainty statement in REFPROP is as follows.

The uncertainties of the equation of state are 0.2% in density, 2% in heat capacity, and 1.5% in the speed of sound, except in the critical region (Tables 10.1, 10.2, 10.3 and 10.4).

J.W. Leachman et al., *Thermodynamic Properties of Cryogenic Fluids*,
International Cryogenics Monograph Series, DOI 10.1007/978-3-319-57835-4_10

Table 10.1 Fixed point properties and correlation limits for fluorine

	Temperature (K)	Pressure (kPa)	Density (mol/dm^3)
Critical point	144.414	5172.4	15.603
Triple Point	53.4811	0.23881	44.917
Upper limit	300	20,000	45.47
Lower limit	53.4811		

Table 10.2 Fluid constants and normal boiling point for fluorine

Molar mass	37.99681 g/mol
Ideal-gas constant	8.31448 J/(mol K)
Pitzer's acentric factor	0.0449
Normal boiling point	85.0368 K

Table 10.3 Fundamental equation of state coefficients and exponents for fluorine

i	N_i	t_i	d_i	p_i	ζ_i
1	1.511447497	0	1		
2	−2.986662884	0.5	1		
3	3.296449051	1.5	1		
4	−2.984586242	2	1		
5	−2.286889665	0.5	2		
6	−1.094921934	1	2		
7	3.047752776	0.5	3		
8	0.115689564	2	3		
9	−1.161001716	0.5	4		
10	0.295656394	1	4		
11	0.071148254	0	5		
12	−0.001713638	0.5	8		
13	0.000665318	0	9		
14	5.060266763	1	2	2	1.07810258
15	−6.292684354	3	2	2	1.07810258
16	6.177848087	4	2	2	1.07810258
17	−1.553661918	5	2	2	1.07810258
18	−2.871706873	1	3	2	1.07810258
19	3.172144805	4	3	2	1.07810258
20	−2.679690252	5	3	2	1.07810258
21	2.718654793	1	4	2	1.07810258
22	−1.07191065	3	4	2	1.07810258
23	1.265973423	5	4	2	1.07810258

(continued)

Table 10.3 (continued)

i	N_i	t_i	d_i	p_i	ζ_i
24	−0.706244695	4	5	2	1.07810258
25	0.268707889	4	6	2	1.07810258
26	0.052725119	1	7	2	1.07810258
27	0.054441148	1	8	2	1.07810258
28	0.00022895	5	12	2	1.07810258
29	−0.0000000000547908264304	30	4	2	2.15620515
30	−0.096427322	20	6	2	3.23430773
31	0.000368084	25	6	2	3.23430773

Table 10.4 Ancillary function coefficients and exponents for fluorine

	α^0	
i	a_i	k_i
0	3.5011231	
1	−3.5816408414	
2	3.5221368165	
3	$-6.0936946 \times 10^{-5}$	4
4	6.319669×10^{-4}	3
5	$-7.4069617 \times 10^{-5}$	−2
6	1.012767	1286.12

$m = 5$, $n = 6$

	Vapor pressure		Saturated liquid density		Saturated vapor density	
i	N_i	k_i	N_i	k_i	N_i	k_i
1	−3.7061	1	0.21286	0.228	−3.6218	0.454
2	−81.517	1.5	4.4011	0.58	−55.363	2.3
3	137.43	1.61	−5.3959	0.908	122.14	2.9
4	−58.617	1.77	4.1347	1.24	−230.92	4
5	−1.3528	7.3	−0.97544	1.6	−338.61	6
6					432.18	5.3

10.2 Property Plots for Fluorine

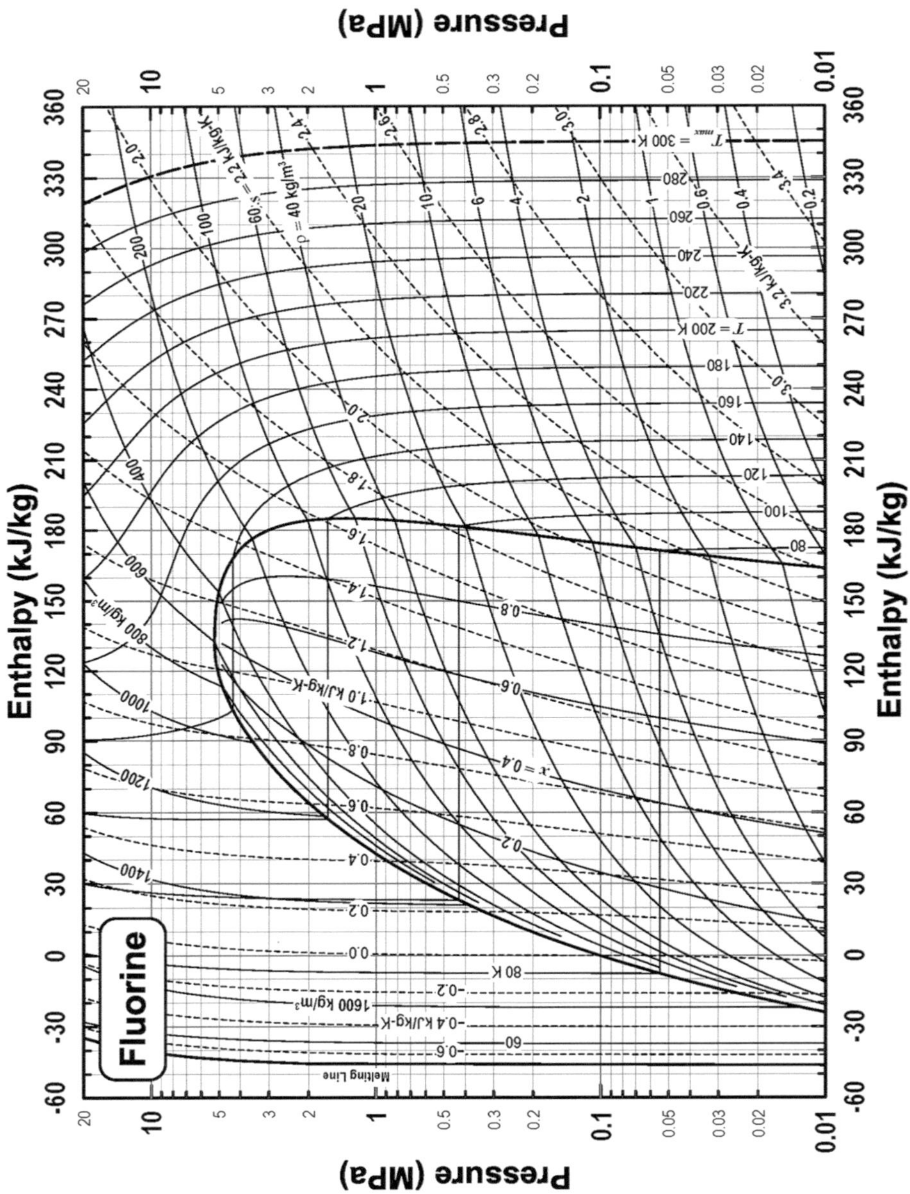
Fluorine
Enthalpy (kJ/kg)
Pressure (MPa)
Melting Line
T_{max} = 300 K
T = 200 K
x = 0.4
s = 2.2 kJ/kg-K
ρ = 40 kg/m³
1600 kg/m³
80 K

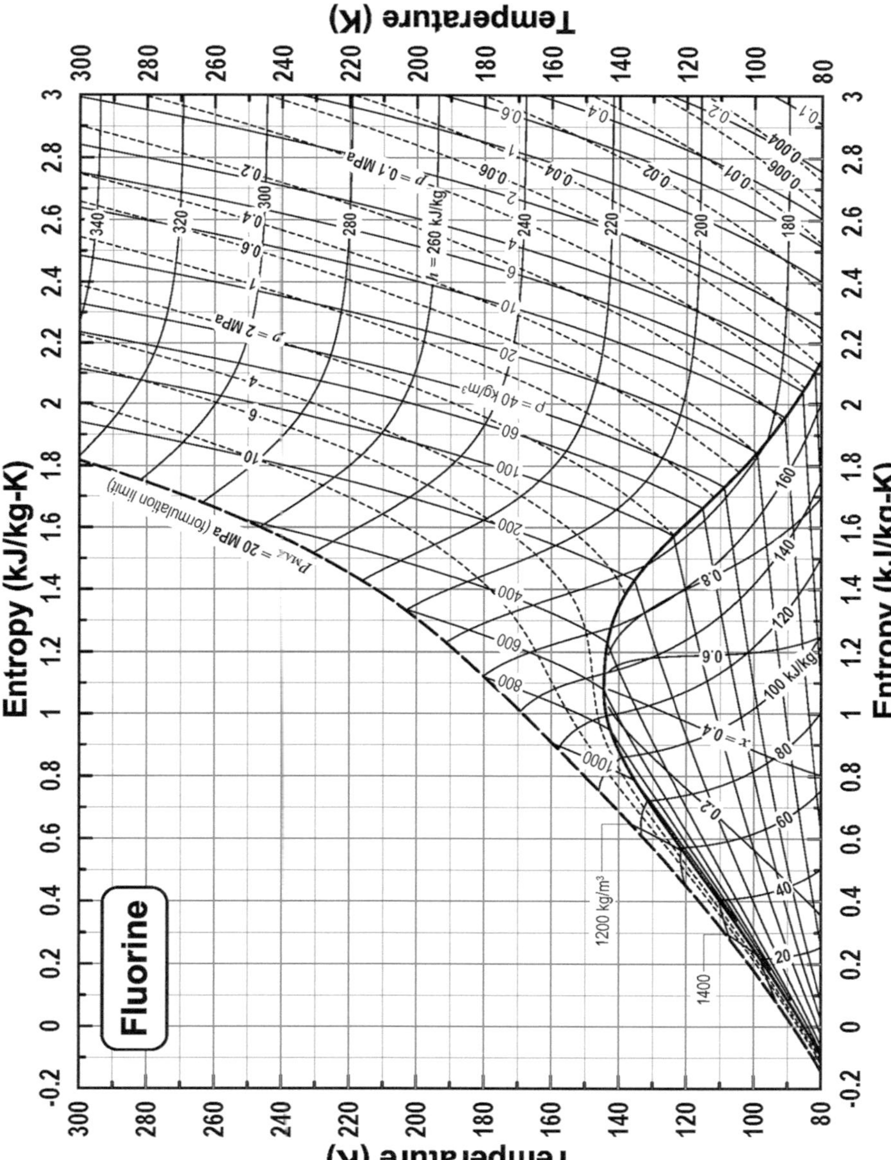
Fluorine
Entropy (kJ/kg-K)
Temperature (K)
p_{MAX} = 20 MPa (formulation limit)
p = 0.1 MPa
p = 2 MPa
h = 260 kJ/kg
ρ = 40 kg/m³
1200 kg/m³
1400
x = 0.4
100 kJ/kg

10.3 Thermodynamic Data Tables for Fluorine

See Tables 10.5 and 10.6.

Table 10.5 Thermodynamic properties of saturated fluorine

Temperature (K)	Pressure (MPa)	Liquid density (kg/m^3)	Vapor density (kg/m^3)	Liquid enthalpy (kJ/kg)	Vapor enthalpy (kJ/kg)	Liquid entropy (kJ/kg K)	Vapor entropy (kJ/kg K)	Liquid C_v (kJ/kg K)	Vapor C_v (kJ/kg K)	Liquid sound speed (m/s)	Vapor sound speed (m/s)
53.481	0.000239	1706.7	0.020414	−46.801	152.27	−0.68618	3.036	0.96855	0.54742	1041.6	127.97
56	0.000512	1690.4	0.041785	−43.078	154.18	−0.61813	2.9044	0.91807	0.54763	1041.6	130.93
60	0.001487	1665.4	0.11345	−37.242	157.21	−0.51749	2.7233	0.90149	0.54823	1043.5	135.46
64	0.003736	1640.3	0.26761	−31.357	160.19	−0.42257	2.5704	0.88282	0.5493	1026.4	139.78
68	0.008333	1615.1	0.56325	−25.445	163.12	−0.33302	2.44	0.86781	0.55106	992.96	143.89
72	0.016849	1589.4	1.08	−19.517	165.97	−0.24839	2.3278	0.85633	0.55368	952.55	147.74
76	0.031404	1563.2	1.9177	−13.569	168.71	−0.16811	2.2302	0.84594	0.55735	911.14	151.34
80	0.054668	1536.5	3.1953	−7.5901	171.31	−0.09163	2.1446	0.83525	0.56217	871.16	154.65
84	0.089831	1509	5.0486	−1.5691	173.76	−0.01848	2.0688	0.82393	0.56818	833.13	157.64
85.037	0.10132	1501.8	5.6413	0	174.37	0	2.0505	0.82091	0.56993	823.56	158.37
88	0.14054	1480.8	7.6295	4.5071	176.03	0.051794	2.0009	0.81217	0.57535	796.75	160.31
92	0.21081	1451.6	11.105	10.652	178.11	0.11955	1.9397	0.80028	0.58358	761.5	162.63
96	0.30496	1421.4	15.658	16.88	179.96	0.18511	1.8839	0.78856	0.59274	726.83	164.57
100	0.42751	1390	21.494	23.208	181.58	0.2488	1.8325	0.77723	0.60268	692.27	166.12
104	0.58314	1357.3	28.846	29.653	182.94	0.31088	1.7848	0.76646	0.61327	657.41	167.25
108	0.77664	1323	37.991	36.238	184	0.37165	1.7398	0.75639	0.62448	621.85	167.94
112	1.0129	1286.9	49.269	42.993	184.73	0.43141	1.6969	0.74722	0.63642	585.17	168.18
116	1.297	1248.6	63.125	49.956	185.06	0.49053	1.6552	0.73923	0.64941	546.91	167.95
120	1.6342	1207.4	80.165	57.183	184.9	0.54944	1.6138	0.73287	0.66412	506.54	167.26
124	2.0301	1162.5	101.27	64.753	184.14	0.60875	1.5715	0.72895	0.68172	463.37	166.15
128	2.4907	1112.5	127.78	72.792	182.57	0.66933	1.527	0.72902	0.70431	416.55	164.65
132	3.023	1054.7	161.98	81.512	179.9	0.73261	1.478	0.73621	0.73576	364.86	162.78
136	3.6347	984.6	208.28	91.311	175.54	0.80125	1.4206	0.75786	0.78411	306.48	160.35
140	4.3357	890.17	278.28	103.17	168.02	0.8817	1.3449	0.81489	0.87069	238.35	156.46
144.14	5.177	681.47	484.3	124.86	144.2	1.0266	1.1608				

Table 10.6 Thermodynamic properties of fluorine

Temperature (K)	Pressure (MPa)	Density (kg/m^3)	Internal energy (kJ/kg)	Enthalpy (kJ/kg)	Entropy (kJ/kg K)	C_v (kJ/kg K)	C_p (kJ/kg K)	Sound speed (m/s)
0.1 MPa								
53.491	0.1	1706.8	−46.798	−46.74	−0.68613	0.96654	1.525	1044.4
60	0.1	1665.5	−37.256	−37.196	−0.51772	0.90068	1.4653	1044.9
70	0.1	1602.4	−22.506	−22.444	−0.29034	0.86179	1.4808	973.64
80	0.1	1536.6	−7.6361	−7.571	−0.09176	0.83532	1.4963	871.36
84.922	0.1	1502.6	−0.24033	−0.17378	−0.00203	0.82125	1.5102	824.62
84.922	0.1	5.5732	156.36	174.3	2.0525	0.56973	0.82467	158.29
90	0.1	5.2263	159.32	178.46	2.1	0.56539	0.81338	163.49
100	0.1	4.664	165.07	186.51	2.1849	0.55973	0.79863	173.11
110	0.1	4.2171	170.74	194.45	2.2606	0.55632	0.78964	182.08
120	0.1	3.8516	176.35	202.31	2.329	0.55429	0.78396	190.54
130	0.1	3.5463	181.94	210.14	2.3916	0.55322	0.7804	198.57
140	0.1	3.2869	187.5	217.93	2.4494	0.55291	0.77832	206.23
150	0.1	3.0536	193.06	225.71	2.503	0.55326	0.7774	213.56
160	0.1	2.8592	198.63	233.48	2.5532	0.55421	0.7774	220.61
170	0.1	2.6982	204.19	241.26	2.6003	0.55573	0.77819	227.39
180	0.1	2.5467	209.78	249.04	2.6449	0.55777	0.77967	233.92
190	0.1	2.4114	215.38	256.85	2.6871	0.56028	0.78174	240.23
200	0.1	2.2899	221.01	264.68	2.7272	0.56323	0.78434	246.33
210	0.1	2.1802	226.67	272.54	2.7656	0.56657	0.7874	252.25
220	0.1	2.0805	232.36	280.43	2.8023	0.57024	0.79083	257.99
230	0.1	1.9896	238.09	288.36	2.8375	0.57418	0.79459	263.56
240	0.1	1.9063	243.86	296.32	2.8714	0.57836	0.79861	268.99
250	0.1	1.8297	249.68	304.33	2.9041	0.58272	0.80284	274.27
260	0.1	1.7591	255.53	312.38	2.9357	0.58722	0.80723	279.43
270	0.1	1.6937	261.43	320.48	2.9662	0.59182	0.81173	284.47
280	0.1	1.6331	267.38	328.62	2.9958	0.59648	0.8163	289.4
290	0.1	1.5766	273.37	336.8	3.0245	0.60117	0.82092	294.23
300	0.1	1.5239	279.41	345.03	3.0525	0.60586	0.82555	298.96
0.2 MPa								
53.501	0.2	1706.9	−46.796	−46.679	−0.68607	0.96455	1.5246	1047.2
60	0.2	1665.7	−37.27	−37.149	−0.51794	0.89987	1.4652	1046.4
70	0.2	1602.6	−22.524	−22.399	−0.29059	0.86179	1.4806	974.19
80	0.2	1536.8	−7.659	−7.5289	−0.09205	0.83547	1.4959	871.8
90	0.2	1466.4	7.4438	7.5802	0.085871	0.80626	1.5294	779.16
91.459	0.2	1455.6	9.6785	9.8159	0.11051	0.80189	1.5361	766.22
91.459	0.2	10.575	158.93	177.84	1.9476	0.58241	0.86093	162.34
100	0.2	9.5301	164.08	185.06	2.0232	0.57229	0.83354	171.11
110	0.2	8.5653	169.95	193.29	2.1016	0.56506	0.814	180.56
120	0.2	7.7935	175.71	201.37	2.1719	0.56061	0.80176	189.35
130	0.2	7.1565	181.4	209.34	2.2358	0.55792	0.79384	197.62
140	0.2	6.6204	187.04	217.25	2.2944	0.55647	0.78873	205.47
150	0.2	6.1619	192.67	225.12	2.3487	0.55599	0.78562	212.95
160	0.2	5.7647	198.28	232.97	2.3993	0.55634	0.784	220.1
170	0.2	5.4168	203.89	240.81	2.4468	0.55739	0.78357	226.97
180	0.2	5.1093	209.5	248.65	2.4916	0.55907	0.78411	233.58

(continued)

Table 10.6 (continued)

Temperature (K)	Pressure (MPa)	Density (kg/m^3)	Internal energy (kJ/kg)	Enthalpy (kJ/kg)	Entropy (kJ/kg K)	C_v (kJ/kg K)	C_p (kJ/kg K)	Sound speed (m/s)
190	0.2	4.8355	215.13	256.49	2.5341	0.56131	0.78546	239.95
200	0.2	4.5899	220.78	264.36	2.5744	0.56405	0.78748	246.11
210	0.2	4.3684	226.46	272.24	2.6129	0.56721	0.79007	252.07
220	0.2	4.1675	232.17	280.16	2.6497	0.57074	0.79313	257.84
230	0.2	3.9844	237.91	288.11	2.685	0.57458	0.79659	263.45
240	0.2	3.8169	243.69	296.09	2.719	0.57867	0.80036	268.91
250	0.2	3.6629	249.52	304.12	2.7518	0.58296	0.80438	274.22
260	0.2	3.521	255.38	312.18	2.7834	0.5874	0.80859	279.4
270	0.2	3.3898	261.29	320.29	2.814	0.59195	0.81294	284.46
280	0.2	3.268	267.24	328.44	2.8436	0.59657	0.81739	289.41
290	0.2	3.1547	273.24	336.64	2.8724	0.60122	0.8219	294.25
300	0.2	3.049	279.28	344.88	2.9003	0.60589	0.82644	299
0.5 MPa								
53.53	0.5	1707.1	−46.787	−46.495	−0.68592	0.95871	1.5232	1055.3
60	0.5	1666.1	−37.31	−37.01	−0.51861	0.89751	1.4649	1050.7
70	0.5	1603.2	−22.576	−22.264	−0.29134	0.86182	1.48	975.83
80	0.5	1537.5	−7.7277	−7.4025	−0.09291	0.83593	1.4948	873.11
90	0.5	1467.3	7.3519	7.6927	0.084848	0.80669	1.5276	780.73
100	0.5	1390.3	22.87	23.229	0.24849	0.77732	1.5847	692.76
101.98	0.5	1374	26.013	26.377	0.27966	0.77183	1.6002	675.1
101.98	0.5	24.922	162.22	182.29	1.8085	0.60783	0.94587	166.73
110	0.5	22.53	167.47	189.66	1.8782	0.59171	0.89667	175.61
120	0.5	20.223	173.7	198.43	1.9545	0.57941	0.85971	185.57
130	0.5	18.405	179.73	206.9	2.0223	0.57177	0.83664	194.66
140	0.5	16.92	185.64	215.19	2.0837	0.56695	0.82145	203.1
150	0.5	15.677	191.45	223.35	2.14	0.56407	0.81127	211.04
160	0.5	14.617	197.22	231.42	2.1921	0.56262	0.80451	218.55
170	0.5	13.699	202.95	239.44	2.2408	0.56233	0.80023	225.7
180	0.5	12.896	208.66	247.43	2.2864	0.56297	0.79784	232.54
190	0.5	12.185	214.37	255.41	2.3295	0.5644	0.7969	239.11
200	0.5	11.551	220.09	263.38	2.3704	0.56649	0.79712	245.42
210	0.5	10.982	225.82	271.35	2.4093	0.56915	0.79828	251.52
220	0.5	10.468	231.58	279.34	2.4465	0.57227	0.80018	257.41
230	0.5	10.001	237.36	287.36	2.4821	0.57578	0.80269	263.12
240	0.5	9.5747	243.18	295.4	2.5164	0.5796	0.80568	268.66
250	0.5	9.1838	249.03	303.47	2.5493	0.58367	0.80906	274.05
260	0.5	8.8241	254.92	311.58	2.5811	0.58794	0.81273	279.3
270	0.5	8.4919	260.85	319.73	2.6119	0.59234	0.81663	284.42
280	0.5	8.1841	266.82	327.91	2.6416	0.59684	0.8207	289.42
290	0.5	7.8981	272.84	336.14	2.6705	0.60139	0.82488	294.32
300	0.5	7.6316	278.9	344.41	2.6985	0.60597	0.82914	299.11
1.0 MPa								
53.579	1	1707.5	−46.774	−46.188	−0.68567	0.9494	1.521	1068.5
60	1	1666.8	−37.377	−36.777	−0.51974	0.89378	1.4644	1057.6
70	1	1604.1	−22.663	−22.039	−0.29258	0.86193	1.4789	978.47
80	1	1538.7	−7.8413	−7.1914	−0.09433	0.83672	1.4931	875.26

(continued)

Table 10.6 (continued)

Temperature (K)	Pressure (MPa)	Density (kg/m^3)	Internal energy (kJ/kg)	Enthalpy (kJ/kg)	Entropy (kJ/kg K)	C_v (kJ/kg K)	C_p (kJ/kg K)	Sound speed (m/s)
90	1	1468.9	7.2001	7.8809	0.083155	0.80741	1.5247	783.32
100	1	1392.4	22.661	23.379	0.2464	0.77792	1.5797	696.1
110	1	1305.9	38.845	39.61	0.40102	0.75178	1.6767	604.69
111.8	1	1288.8	41.875	42.651	0.42844	0.74766	1.7023	587.03
111.8	1	48.647	164.14	184.7	1.699	0.6358	1.0725	168.18
120	1	43.464	170.08	193.08	1.7714	0.61172	0.98174	178.45
130	1	38.8	176.79	202.56	1.8473	0.59457	0.92006	189.28
140	1	35.222	183.17	211.56	1.914	0.58398	0.88266	198.91
150	1	32.35	189.35	220.26	1.974	0.57718	0.85811	207.71
160	1	29.971	195.39	228.75	2.0289	0.57289	0.8414	215.88
170	1	27.956	201.34	237.11	2.0795	0.57044	0.82989	223.54
180	1	26.221	207.23	245.36	2.1267	0.56942	0.82208	230.78
190	1	24.705	213.08	253.56	2.171	0.56955	0.81699	237.68
200	1	23.367	218.91	261.71	2.2128	0.57061	0.81397	244.27
210	1	22.176	224.75	269.84	2.2525	0.57243	0.81255	250.59
220	1	21.106	230.58	277.96	2.2903	0.57488	0.81239	256.68
230	1	20.139	236.44	286.09	2.3264	0.57784	0.81322	262.56
240	1	19.26	242.31	294.23	2.3611	0.58122	0.81484	268.24
250	1	18.458	248.21	302.39	2.3944	0.58493	0.81708	273.76
260	1	17.721	254.15	310.57	2.4265	0.58889	0.81981	279.12
270	1	17.043	260.11	318.79	2.4575	0.59304	0.82292	284.35
280	1	16.416	266.12	327.03	2.4875	0.59733	0.82631	289.44
290	1	15.835	272.16	335.32	2.5165	0.60172	0.82993	294.41
300	1	15.294	278.25	343.63	2.5447	0.60616	0.8337	299.28
2.0 MPa								
53.677	2	1708.2	−46.746	−45.575	−0.68516	0.93231	1.5168	1093.2
60	2	1668.3	−37.51	−36.311	−0.52197	0.88705	1.4635	1070.7
70	2	1605.8	−22.834	−21.589	−0.29505	0.86238	1.4769	983.51
80	2	1541	−8.0657	−6.7679	−0.09716	0.83836	1.4896	879.43
90	2	1471.9	6.9012	8.26	0.079811	0.80884	1.5191	788.41
100	2	1396.5	22.253	23.685	0.24228	0.77911	1.5702	702.67
110	2	1311.9	38.258	39.782	0.39564	0.75273	1.6579	613.64
120	2	1211	55.488	57.14	0.54656	0.73274	1.8397	511.65
123.72	2	1165.8	62.488	64.204	0.60452	0.72912	1.9712	466.54
123.72	2	99.617	164.14	184.21	1.5746	0.68034	1.3984	166.24
130	2	88.954	169.8	192.28	1.6382	0.6464	1.1969	176.51
140	2	77.529	177.63	203.43	1.7209	0.61861	1.0511	189.66
150	2	69.488	184.75	213.53	1.7906	0.60294	0.97555	200.68
160	2	63.339	191.46	223.03	1.852	0.59295	0.9291	210.39
170	2	58.405	197.92	232.16	1.9073	0.58641	0.89811	219.2
180	2	54.316	204.21	241.03	1.958	0.58226	0.87661	227.32
190	2	50.848	210.38	249.71	2.005	0.57991	0.86145	234.9
200	2	47.854	216.48	258.27	2.0489	0.57898	0.85079	242.05
210	2	45.232	222.52	266.74	2.0902	0.57919	0.84343	248.84
220	2	42.912	228.54	275.15	2.1293	0.58032	0.83857	255.31
230	2	40.839	234.55	283.52	2.1665	0.5822	0.83564	261.52

(continued)

Table 10.6 (continued)

Temperature (K)	Pressure (MPa)	Density (kg/m³)	Internal energy (kJ/kg)	Enthalpy (kJ/kg)	Entropy (kJ/kg K)	C_v (kJ/kg K)	C_p (kJ/kg K)	Sound speed (m/s)
240	2	38.972	240.55	291.87	2.202	0.58468	0.8342	267.49
250	2	37.281	246.56	300.21	2.2361	0.58764	0.83394	273.25
260	2	35.738	252.59	308.55	2.2688	0.59099	0.83459	278.83
270	2	34.326	258.64	316.9	2.3003	0.59463	0.83597	284.24
280	2	33.026	264.71	325.27	2.3307	0.59849	0.83792	289.5
290	2	31.825	270.82	333.66	2.3602	0.60252	0.8403	294.62
300	2	30.712	276.96	342.08	2.3887	0.60665	0.84303	299.62
5.0 MPa								
53.968	5	1710.3	−46.664	−43.741	−0.68368	0.89141	1.5055	1156.1
60	5	1672.5	−37.903	−34.914	−0.52861	0.87199	1.4612	1105.1
70	5	1611.1	−23.336	−20.232	−0.30231	0.86531	1.4712	996.78
80	5	1547.8	−8.7174	−5.487	−0.10543	0.8436	1.4799	891.23
90	5	1480.8	6.0403	9.417	0.070089	0.81296	1.5036	803.27
100	5	1408.3	21.092	24.642	0.23046	0.78234	1.5447	721.7
110	5	1328.4	36.629	40.393	0.38054	0.75559	1.6111	638.53
120	5	1236.9	52.993	57.035	0.52526	0.73379	1.7306	548.09
130	5	1122.9	71.039	75.492	0.67286	0.72097	2.0063	441.41
140	5	937.88	94.506	99.838	0.85275	0.76044	3.3244	286.4
143.33	5	755.34	111.14	117.76	0.97893	0.95519	16.589	169.23
143.33	5	398.57	141.46	154.01	1.2318	1.041	26.833	148.4
150	5	247.35	163.86	184.07	1.4384	0.71991	2.2223	177.58
160	5	197.64	176.21	201.51	1.5512	0.65893	1.4615	195.31
170	5	171	185.56	214.8	1.6318	0.63491	1.228	208.41
180	5	152.96	193.75	226.44	1.6983	0.62088	1.1097	219.36
190	5	139.48	201.3	237.15	1.7563	0.61154	1.0376	228.98
200	5	128.81	208.45	247.27	1.8082	0.60512	0.98937	237.67
210	5	120.04	215.33	256.98	1.8556	0.60083	0.95542	245.66
220	5	112.65	222.02	266.41	1.8994	0.59819	0.93086	253.1
230	5	106.3	228.58	275.62	1.9404	0.59689	0.91284	260.1
240	5	100.74	235.05	284.68	1.979	0.59668	0.89958	266.74
250	5	95.831	241.45	293.62	2.0155	0.59736	0.88989	273.06
260	5	91.443	247.81	302.49	2.0502	0.59877	0.88291	279.13
270	5	87.488	254.14	311.29	2.0835	0.60076	0.87804	284.96
280	5	83.9	260.46	320.05	2.1153	0.60323	0.87483	290.6
290	5	80.624	266.77	328.79	2.146	0.60607	0.87291	296.06
300	5	77.617	273.09	337.51	2.1756	0.6092	0.87202	301.36
10.0 MPa								
54.45	10	1713.2	−46.527	−40.69	−0.6813	0.84973	1.4912	1231.3
60	10	1679.1	−38.535	−32.579	−0.53943	0.86	1.4585	1148.3
70	10	1619.4	−24.13	−17.955	−0.31399	0.87393	1.4629	1014.3
80	10	1558.5	−9.7374	−3.321	−0.11859	0.85252	1.4656	909.38
90	10	1494.5	4.7113	11.402	0.054807	0.8185	1.4813	827.43
100	10	1426.3	19.338	26.349	0.21226	0.78578	1.5102	752.43
110	10	1352.6	34.262	41.656	0.35811	0.75855	1.5539	676.68
120	10	1271.6	49.64	57.504	0.49597	0.73673	1.6213	597.35
130	10	1179.3	65.76	74.24	0.62987	0.71991	1.7374	512.6

(continued)

Table 10.6 (continued)

Temperature (K)	Pressure (MPa)	Density (kg/m^3)	Internal energy (kJ/kg)	Enthalpy (kJ/kg)	Entropy (kJ/kg K)	C_v (kJ/kg K)	C_p (kJ/kg K)	Sound speed (m/s)
140	10	1067.6	83.249	92.616	0.76595	0.70965	1.9656	420.87
150	10	917.3	103.56	114.46	0.91646	0.71296	2.4743	324.06
160	10	694.96	129.52	143.91	1.1062	0.73296	3.409	240.59
170	10	488.69	154.57	175.03	1.2951	0.70379	2.5705	217.1
180	10	386.33	170.73	196.61	1.4187	0.67381	1.8459	221.98
190	10	328.55	182.76	213.2	1.5085	0.65648	1.5082	230.89
200	10	290.48	192.84	227.27	1.5806	0.64451	1.3213	240.08
210	10	262.9	201.82	239.85	1.6421	0.63541	1.2044	248.79
220	10	241.64	210.1	251.48	1.6962	0.62836	1.1253	256.92
230	10	224.54	217.9	262.44	1.7449	0.62298	1.069	264.52
240	10	210.35	225.37	272.91	1.7895	0.61902	1.0276	271.65
250	10	198.3	232.59	283.02	1.8308	0.61632	0.9965	278.39
260	10	187.87	239.64	292.86	1.8694	0.61469	0.97279	284.78
270	10	178.72	246.54	302.5	1.9057	0.61399	0.95461	290.89
280	10	170.59	253.35	311.97	1.9402	0.61407	0.94064	296.76
290	10	163.3	260.08	321.32	1.973	0.61482	0.92992	302.41
300	10	156.71	266.76	330.58	2.0044	0.61612	0.92178	307.87
20.0 MPa								
55.399	20	1718.3	−46.245	−34.605	−0.67662	0.82949	1.4753	1305.1
60	20	1691.4	−39.712	−27.887	−0.56012	0.86661	1.4565	1198.8
70	20	1635	−25.575	−13.342	−0.33589	0.89762	1.4497	1039.4
80	20	1578.2	−11.566	1.1064	−0.14294	0.86699	1.4419	944.3
90	20	1519.1	2.3722	15.538	0.027031	0.8222	1.4464	877.18
100	20	1457.1	16.341	30.067	0.18009	0.7833	1.4605	815.2
110	20	1392	30.4	44.769	0.3202	0.75402	1.4808	750.77
120	20	1323.4	44.588	59.701	0.45011	0.7327	1.5068	683.46
130	20	1250.6	58.941	74.933	0.57201	0.71716	1.5413	614.72
140	20	1172.8	73.517	90.571	0.68788	0.70589	1.5888	546.27
150	20	1088.4	88.392	106.77	0.7996	0.6979	1.6535	480.5
160	20	996.61	103.62	123.69	0.9088	0.69229	1.7329	421.11
170	20	898.24	119.15	141.42	1.0162	0.68767	1.8088	372.37
180	20	798.04	134.65	159.71	1.1208	0.68249	1.8375	337.08
190	20	704.63	149.47	177.85	1.2189	0.67642	1.7754	315.29
200	20	625.34	163.02	195.01	1.3069	0.67043	1.6504	304.19
210	20	561.27	175.22	210.86	1.3843	0.66496	1.5224	299.81
220	20	509.8	186.29	225.53	1.4525	0.65982	1.4146	299.26
230	20	467.97	196.48	239.22	1.5134	0.65481	1.3268	300.96
240	20	433.47	205.97	252.11	1.5683	0.64998	1.2548	304.04
250	20	404.6	214.92	264.35	1.6183	0.64549	1.1954	307.96
260	20	380.08	223.44	276.06	1.6642	0.64151	1.1463	312.39
270	20	359	231.6	287.31	1.7067	0.63813	1.1058	317.09
280	20	340.65	239.48	298.2	1.7463	0.63542	1.0724	321.93
290	20	324.5	247.14	308.78	1.7834	0.63339	1.0448	326.81
300	20	310.15	254.62	319.11	1.8184	0.63202	1.0222	331.68

Reference

1. K.M. de Reuck, *International thermodynamic tables of the fluid state-11 fluorine* (Pergamon Press, International Union of Pure and Applied Chemistry, Oxford, 1990)

Chapter 11
Argon

11.1 Formulation for Argon

The formulation for argon was published by Tegeler et al. [1]. The uncertainty statement in REFPROP is as follows.

The estimated uncertainty in density is less than 0.02% for pressures up to 12 MPa and temperatures up to 340 K with the exception of the critical region and less than 0.03% for pressures up to 30 MPa and temperatures between 235 and 520 K. Elsewhere, the uncertainty in density is generally within 0.2%. In the region with densities up to half the critical density and for temperatures between 90 and 450 K, the estimated uncertainty of calculated speeds of sound is in general less than 0.02%. In the liquid and supercritical regions, the uncertainty is less than 1%. The uncertainty in heat capacities is within 0.3% for the vapor and 2% for the liquid. The formulation gives reasonable extrapolation behavior up to very high pressures (50 GPa) and temperatures (17,000 K) (Tables 11.1, 11.2, 11.3 and 11.4).

J.W. Leachman et al., *Thermodynamic Properties of Cryogenic Fluids*,
International Cryogenics Monograph Series, DOI 10.1007/978-3-319-57835-4_11

Table 11.1 Fixed point properties and correlation limits for argon

	Temperature (K)	Pressure (kPa)	Density (mol/dm^3)
Critical point	150.687	4863.0	13.4074
Triple point	83.8058	68.891	35.465
Upper limit	2000	1,000,000	50.65
Lower limit	83.8058		

Table 11.2 Fluid constants and normal boiling point for argon

Molar mass	39.948 g/mol
Ideal-gas constant	8.31451 J/(mol K)
Pitzer's acentric factor	−0.00219
Normal boiling point	87.302 K

Table 11.3 Fundamental equation of state coefficients and exponents for argon

i	N_i	t_i	d_i	p_i	φ_i	β_i	γ_i	D_i
1	0.088722305	0	1					
2	0.705148052	0.25	1					
3	−1.682011565	1	1					
4	−0.149090144	2.75	1					
5	−0.120248046	4	1					
6	−0.121649788	0	2					
7	0.400359336	0.25	2					
8	−0.271360627	0.75	2					
9	0.242119246	2.75	2					
10	0.005788958	0	3					
11	−0.041097336	2	3					
12	0.024710762	0.75	4					
13	−0.321813918	3	1	1				
14	0.332300177	3.5	1	1				
15	0.031019986	1	3	1				
16	−0.030777086	2	4	1				
17	0.093891137	4	4	1				
18	−0.090643211	3	5	1				
19	−0.000457783	0	7	1				
20	-8.26597×10^{-5}	0.5	10	1				
21	0.000130134	1	10	1				
22	−0.01139784	1	2	2				
23	−0.02445517	7	2	2				
24	−0.064324067	5	4	2				
25	0.058889471	6	4	2				

(continued)

Table 11.3 (continued)

i	N_i	t_i	d_i	p_i	φ_i	β_i	γ_i	D_i
26	−0.000649336	6	8	2				
27	−0.013889862	10	3	3				
28	0.404898393	13	5	3				
29	−0.386125196	14	5	3				
30	−0.188171423	11	6	3				
31	0.159776476	14	6	3				
32	0.053985519	8	7	3				
33	−0.028953418	14	7	3				
34	−0.013025413	6	8	3				
35	0.00289487	7	9	3				
36	−0.002264713	24	5	4				
37	0.001761646	22	6	4				
38	0.005855245	3	2		20	250	1.11	1
39	−0.692519083	1	1		20	375	1.14	1
40	1.531549003	0	2		20	300	1.17	1
41	−0.002738045	0	3		20	225	1.11	1

Table 11.4 Ancillary function coefficients and exponents for argon

	α^0
i	a_i
0	2.5
1	−10.2938148005
2	−0.0003415238

$m = 0$, $n = 0$

	Vapor pressure		Saturated liquid density		Saturated vapor density	
i	N_i	k_i	N_i	k_i	N_i	k_i
1	−5.94098	1	1.500426	0.334	−2.9182	0.72
2	1.355389	1.5	−0.31381	0.6666	0.09793	1.25
3	−0.46498	2	0.086462	2.3333	−1.3721	0.32
4	−1.5399	4.5	−0.04148	4	−2.2898	4.34

The functional form of the ancillary equations for argon are different than those of the other fluids, see Chap. 2 for the correct form

11.2 Property Plots for Argon

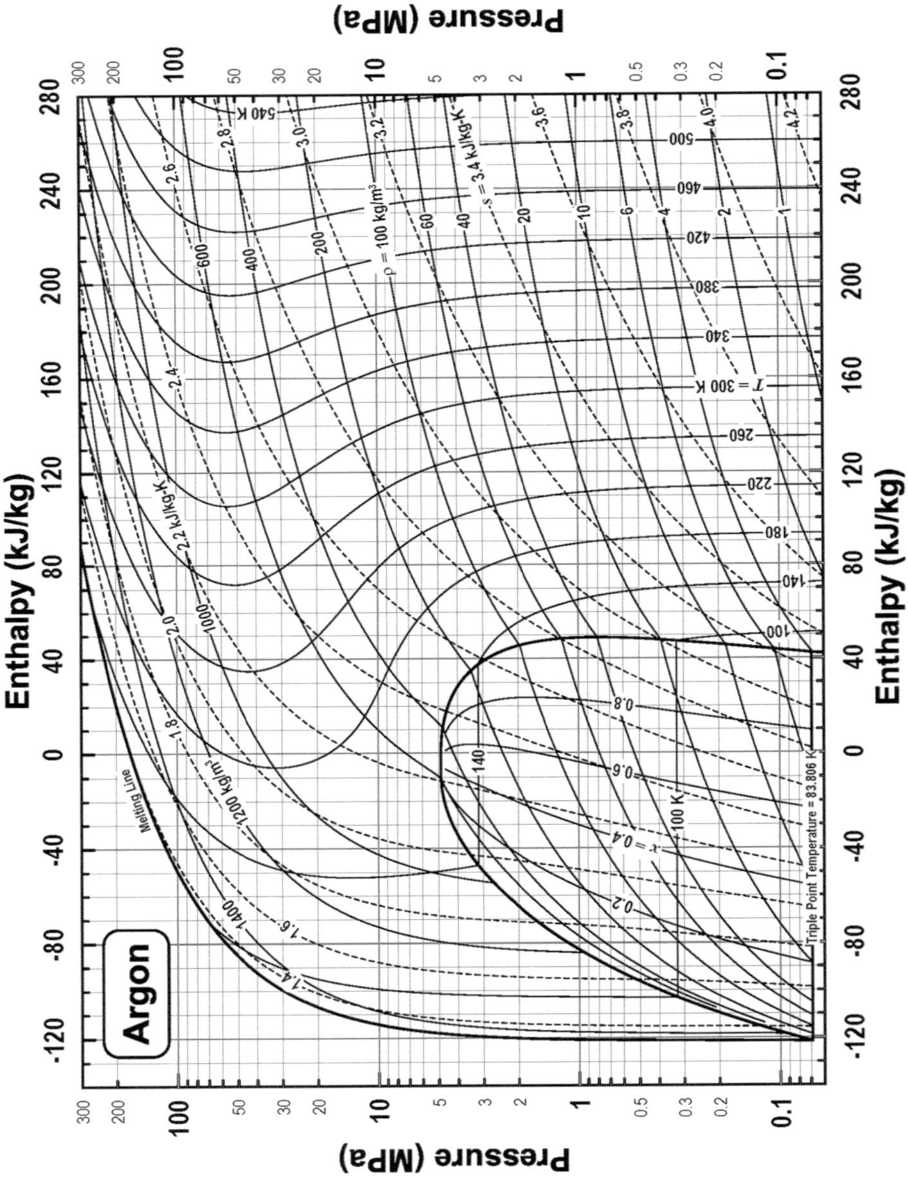
Argon
Enthalpy (kJ/kg)
Pressure (MPa)
Melting Line
Triple Point Temperature = 83.806 K
T = 300 K
ρ = 100 kg/m³
s = 3.4 kJ/kg-K
2.2 kJ/kg-K
1200 kg/m³
x = 0.4

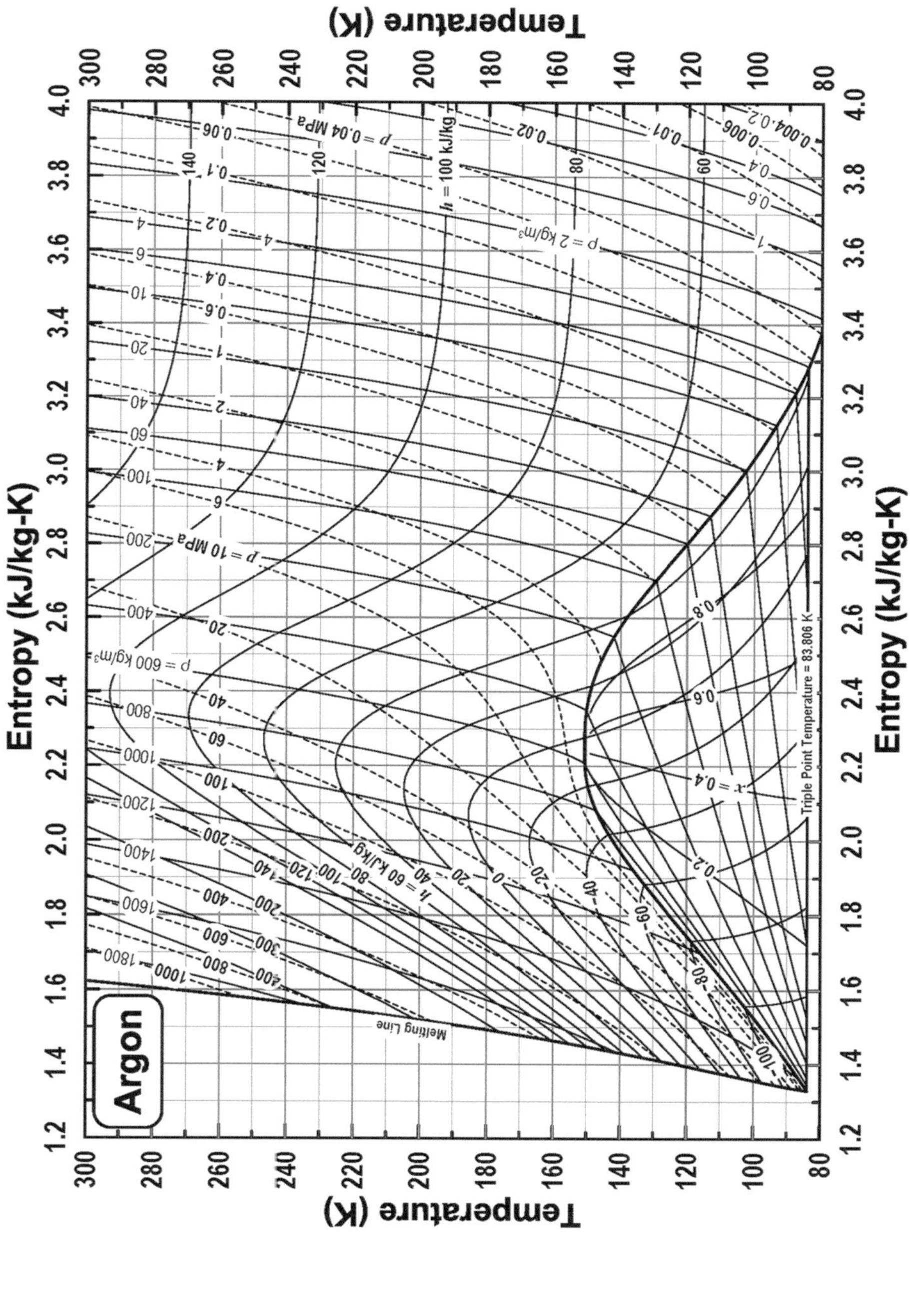
Argon
Entropy (kJ/kg-K)
Temperature (K)
Melting Line
Triple Point Temperature = 83.806 K
p = 10 MPa
p = 0.04 MPa
h = 100 kJ/kg
h = 60 kJ/kg
ρ = 600 kg/m³
ρ = 2 kg/m³
x = 0.4

11.3 Thermodynamic Data Tables for Argon

See Tables 11.5 and 11.6.

Table 11.5 Thermodynamic properties of saturated argon

Temperature (K)	Pressure (MPa)	Liquid density (kg/m^3)	Vapor density (kg/m^3)	Liquid enthalpy (kJ/kg)	Vapor enthalpy (kJ/kg)	Liquid entropy (kJ/kg K)	Vapor entropy (kJ/kg K)	Liquid C_v (kJ/kg K)	Vapor C_v (kJ/kg K)	Liquid sound speed (m/s)	Vapor sound speed (m/s)
83.806	0.068891	1416.8	4.0546	−121.44	42.281	1.3295	3.283	0.5496	0.32471	862.43	168.12
86	0.08811	1403.4	5.0795	−118.98	43.133	1.3583	3.2433	0.54094	0.32674	847.35	169.87
87.302	0.10132	1395.4	5.7736	−117.52	43.618	1.375	3.2208	0.53612	0.32804	838.33	170.86
89	0.12078	1384.9	6.7829	−115.61	44.227	1.3965	3.1924	0.53014	0.32983	826.48	172.12
92	0.16199	1366	8.8829	−112.23	45.231	1.4335	3.1451	0.52031	0.33329	805.28	174.2
95	0.21305	1346.8	11.435	−108.82	46.137	1.4696	3.1007	0.51126	0.33715	783.72	176.11
98	0.27532	1327.1	14.499	−105.38	46.934	1.5048	3.059	0.50287	0.34142	761.78	177.85
101	0.35023	1306.9	18.141	−101.89	47.614	1.5392	3.0195	0.49506	0.34614	739.4	179.41
104	0.4392	1286.2	22.431	−98.365	48.166	1.573	2.9819	0.48777	0.35133	716.54	180.8
107	0.54371	1264.8	27.45	−94.779	48.579	1.6062	2.946	0.48098	0.35703	693.16	182.01
110	0.66526	1242.8	33.287	−91.129	48.84	1.639	2.9114	0.47468	0.36331	669.2	183.03
113	0.80535	1219.9	40.047	−87.404	48.935	1.6713	2.8779	0.46889	0.37022	644.6	183.87
117	1.0237	1188	50.711	−82.301	48.775	1.7141	2.8345	0.46204	0.38057	610.64	184.7
120	1.213	1162.8	60.144	−78.353	48.413	1.7461	2.8025	0.45763	0.38934	584.19	185.09
123	1.4262	1136.3	71.021	−74.282	47.814	1.7781	2.7707	0.45395	0.39915	556.75	185.27
126	1.6648	1108.4	83.59	−70.064	46.947	1.8102	2.7389	0.45116	0.41024	528.16	185.24
129	1.9305	1078.6	98.177	−65.671	45.767	1.8428	2.7067	0.44947	0.42285	498.21	184.98
132	2.2252	1046.5	115.23	−61.062	44.217	1.876	2.6736	0.44921	0.43734	466.63	184.49
135	2.5509	1011.5	135.39	−56.177	42.211	1.9102	2.639	0.45085	0.4544	433.1	183.74
138	2.9096	972.57	159.65	−50.93	39.621	1.946	2.6022	0.45504	0.47581	397.21	182.6
141	3.3037	928.13	189.64	−45.174	36.238	1.9843	2.5617	0.46313	0.50484	358.17	180.81
144	3.7363	874.98	228.48	−38.635	31.672	2.0268	2.515	0.47972	0.54719	313.8	177.93
147	4.2111	805.45	283.86	−30.656	24.977	2.0777	2.4562	0.52174	0.61816	258.79	172.74
150.69	4.863	535.6	535.6	−4.3316	−4.3316	2.2476	2.2476				

Table 11.6 Thermodynamic properties of argon

Temperature (K)	Pressure (MPa)	Density (kg/m³)	Internal energy (kJ/kg)	Enthalpy (kJ/kg)	Entropy (kJ/kg K)	C_v (kJ/kg K)	C_p (kJ/kg K)	Sound speed (m/s)
0.1 MPa								
83.814	0.1	1416.8	−121.48	−121.41	1.3295	0.54961	1.1156	862.52
87.178	0.1	1396.2	−117.73	−117.66	1.3734	0.53657	1.1171	839.2
87.178	0.1	5.7043	26.042	43.573	3.2229	0.32791	0.56541	170.77
90	0.1	5.5077	27.004	45.161	3.2408	0.3257	0.56	173.85
100	0.1	4.9152	30.344	50.689	3.2991	0.32058	0.54702	184.16
110	0.1	4.4436	33.614	56.119	3.3508	0.31778	0.53948	193.76
120	0.1	4.0577	36.844	61.488	3.3975	0.31612	0.53472	202.8
130	0.1	3.7352	40.046	66.818	3.4402	0.31506	0.53153	211.39
140	0.1	3.4513	43.231	72.122	3.4795	0.31436	0.52928	219.6
150	0.1	3.2255	46.403	77.406	3.516	0.31388	0.52764	227.49
160	0.1	3.0202	49.566	82.676	3.55	0.31353	0.52641	235.1
170	0.1	2.8398	52.722	87.935	3.5819	0.31327	0.52546	242.45
180	0.1	2.63	55.873	93.186	3.6119	0.31308	0.52472	249.57
190	0.1	2.5374	59.019	98.43	3.6402	0.31294	0.52412	256.48
200	0.1	2.4093	62.162	103.67	3.6671	0.31282	0.52364	263.21
210	0.1	2.2936	65.303	108.9	3.6926	0.31273	0.52324	269.76
220	0.1	2.1885	68.441	114.13	3.717	0.31266	0.52291	276.15
230	0.1	2.0927	71.577	119.36	3.7402	0.3126	0.52264	282.4
240	0.1	2.005	74.712	124.59	3.7625	0.31255	0.5224	288.5
250	0.1	1.9244	77.845	129.81	3.7838	0.31251	0.5222	294.48
260	0.1	1.85	80.978	135.03	3.8043	0.31248	0.52202	300.33
270	0.1	1.7812	84.109	140.25	3.824	0.31245	0.52187	306.07
280	0.1	1.7174	87.24	145.47	3.8429	0.31242	0.52174	311.7
290	0.1	1.6579	90.369	150.68	3.8612	0.3124	0.52162	317.24
300	0.1	1.6025	93.499	155.9	3.8789	0.31238	0.52152	322.67
400	0.1	1.2012	124.77	208.02	4.0289	0.31229	0.52093	372.65
500	0.1	0.96076	156.01	260.1	4.1451	0.31225	0.52069	416.64
600	0.1	0.80058	187.25	312.16	4.24	0.31224	0.52057	456.4
700	0.1	0.68619	218.48	364.21	4.3202	0.31223	0.5205	492.95
800	0.1	0.60042	249.71	416.26	4.3897	0.31222	0.52045	526.97
900	0.1	0.53371	280.94	468.3	4.451	0.31222	0.52043	558.92
1000	0.1	0.48034	312.16	520.35	4.5059	0.31222	0.52041	589.14
0.2 MPa								
83.839	0.2	1416.9	−121.48	−121.34	1.3295	0.54963	1.1152	862.83
90	0.2	1378.8	−114.6	−114.46	1.4087	0.52685	1.1209	819.82
94.29	0.2	1351.4	−109.78	−109.63	1.4612	0.51334	1.1317	788.86
94.29	0.2	10.787	27.391	45.932	3.111	0.3362	0.59453	175.67
100	0.2	10.074	29.425	49.278	3.1454	0.33023	0.57856	181.93
110	0.2	9.0496	32.867	54.968	3.1997	0.3239	0.56097	192.1
120	0.2	8.2288	36.215	60.52	3.248	0.32029	0.55036	201.53
130	0.2	7.5523	39.505	65.987	3.2918	0.31805	0.54344	210.4
140	0.2	6.9833	42.757	71.396	3.3318	0.31659	0.53867	218.82
150	0.2	6.4959	45.981	76.765	3.3689	0.31559	0.53523	226.87

(continued)

Table 11.6 (continued)

Temperature (K)	Pressure (MPa)	Density (kg/m^3)	Internal energy (kJ/kg)	Enthalpy (kJ/kg)	Entropy (kJ/kg K)	C_v (kJ/kg K)	C_p (kJ/kg K)	Sound speed (m/s)
160	0.2	6.0759	49.187	82.104	3.4033	0.31488	0.53267	234.6
170	0.2	5.7073	52.378	87.42	3.4356	0.31436	0.53072	242.05
180	0.2	5.3819	55.558	92.72	3.4659	0.31397	0.52919	249.26
190	0.2	5.0922	58.729	98.005	3.4944	0.31368	0.52798	256.24
200	0.2	4.8325	61.894	103.28	3.5215	0.31344	0.52699	263.02
210	0.2	4.5985	65.053	108.55	3.5472	0.31326	0.52619	269.62
220	0.2	4.3863	68.208	113.8	3.5717	0.31312	0.52552	276.05
230	0.2	4.193	71.358	119.06	3.595	0.313	0.52496	282.33
240	0.2	4.0163	74.506	124.3	3.6173	0.3129	0.52448	288.47
250	0.2	3.8539	77.651	129.55	3.6387	0.31282	0.52407	294.47
260	0.2	3.7043	80.794	134.79	3.6593	0.31275	0.52372	300.34
270	0.2	3.5659	83.935	140.02	3.679	0.31269	0.52342	306.1
280	0.2	3.4376	87.074	145.25	3.6981	0.31264	0.52315	311.75
290	0.2	3.3183	90.212	150.48	3.7164	0.3126	0.52292	317.3
300	0.2	3.207	93.348	155.71	3.7341	0.31257	0.52271	322.75
400	0.2	2.4024	124.67	207.91	3.8843	0.31238	0.52153	372.8
500	0.2	1.9212	155.94	260.04	4.0006	0.31231	0.52105	416.82
600	0.2	1.6008	187.19	312.13	4.0956	0.31228	0.5208	456.58
700	0.2	1.372	218.43	364.2	4.1759	0.31226	0.52066	493.14
800	0.2	1.2005	249.67	416.27	4.2454	0.31225	0.52057	527.15
900	0.2	1.0671	280.9	468.32	4.3067	0.31224	0.52052	559.1
1000	0.2	0.96045	312.13	520.37	4.3616	0.31223	0.52048	589.31
0.5 MPa								
83.914	0.5	1417.3	−121.47	−121.12	1.3296	0.54969	1.1142	863.76
90	0.5	1379.8	−114.69	−114.33	1.4077	0.52721	1.1194	821.46
100	0.5	1314.4	−103.38	−103	1.5271	0.4978	1.1523	748.14
105.8	0.5	1273.4	−96.612	−96.219	1.593	0.48363	1.1862	702.57
105.8	0.5	25.352	28.709	48.432	2.9602	0.35469	0.67091	181.55
110	0.5	24.05	30.402	51.192	2.9858	0.34659	0.64536	186.57
120	0.5	21.524	34.203	57.434	3.0401	0.33464	0.60684	197.45
130	0.5	19.548	37.803	63.381	3.0877	0.32791	0.58435	207.29
140	0.5	17.944	41.282	69.147	3.1305	0.32373	0.56985	216.41
150	0.5	16.605	44.681	74.793	3.1694	0.32098	0.55987	224.97
160	0.5	15.466	48.024	80.354	3.2053	0.31907	0.55268	233.1
170	0.5	14.482	51.327	85.852	3.2387	0.3177	0.54731	240.87
180	0.5	13.622	54.6	91.304	3.2698	0.31669	0.54318	248.32
190	0.5	12.864	57.85	96.719	3.2991	0.31592	0.53994	255.51
200	0.5	12.188	61.081	102.1	3.3267	0.31533	0.53734	262.47
210	0.5	11.582	64.298	107.47	3.3529	0.31487	0.53523	269.21
220	0.5	11.036	67.503	112.81	3.3777	0.3145	0.53349	275.77
230	0.5	10.54	70.699	118.14	3.4014	0.31419	0.53204	282.15
240	0.5	10.087	73.886	123.45	3.424	0.31395	0.53081	288.37
250	0.5	9.6733	77.066	128.75	3.4457	0.31374	0.52977	294.45
260	0.5	9.2924	80.241	134.05	3.4664	0.31357	0.52887	300.4

(continued)

Table 11.6 (continued)

Temperature (K)	Pressure (MPa)	Density (kg/m³)	Internal energy (kJ/kg)	Enthalpy (kJ/kg)	Entropy (kJ/kg K)	C_v (kJ/kg K)	C_p (kJ/kg K)	Sound speed (m/s)
270	0.5	8.9409	83.41	139.33	3.4864	0.31343	0.52809	306.22
280	0.5	8.6155	86.575	144.61	3.5056	0.31331	0.52742	311.92
290	0.5	8.3133	89.737	149.88	3.5241	0.3132	0.52683	317.51
300	0.5	8.0319	92.895	155.15	3.5419	0.31311	0.52631	323
400	0.5	6.0064	124.36	207.6	3.6929	0.31264	0.52333	373.27
500	0.5	4.8005	155.71	259.87	3.8095	0.31247	0.52211	417.35
600	0.5	3.9991	187.02	312.05	3.9046	0.31239	0.5215	457.14
700	0.5	3.4274	218.29	364.18	3.985	0.31235	0.52115	493.69
800	0.5	2.9989	249.55	416.28	4.0546	0.31232	0.52094	527.69
900	0.5	2.6658	280.81	468.37	4.1159	0.3123	0.52079	559.62
1000	0.5	2.3994	312.05	520.44	4.1708	0.31228	0.52069	589.82
1.0 MPa								
84.039	1	1417.9	−121.46	−120.76	1.3298	0.5498	1.1125	865.3
90	1	1381.3	−114.84	−114.12	1.4061	0.52781	1.1171	824.17
100	1	1316.5	−103.57	−102.81	1.5252	0.49837	1.1483	751.59
110	1	1244.7	−91.846	−91.043	1.6373	0.475	1.2125	672.31
116.6	1	1191.3	−83.661	−82.822	1.7099	0.46268	1.2851	614.12
116.6	1	49.546	28.624	48.807	2.8388	0.37946	0.80069	184.63
120	1	47.202	30.265	51.451	2.8611	0.36817	0.75594	189.35
130	1	41.787	34.64	58.571	2.9182	0.34822	0.67741	201.51
140	1	37.747	38.623	65.115	2.9667	0.33748	0.63497	212.09
150	1	34.551	42.381	71.323	3.0095	0.3309	0.60857	221.67
160	1	31.929	45.995	77.314	3.0482	0.32657	0.59074	230.54
170	1	29.723	49.511	83.155	3.0836	0.32356	0.57801	238.88
180	1	27.833	52.957	88.885	3.1164	0.32138	0.56855	246.79
190	1	26.189	56.349	94.533	3.1469	0.31977	0.5613	254.34
200	1	24.743	59.701	100.12	3.1755	0.31853	0.5556	261.59
210	1	23.458	63.02	105.65	3.2025	0.31757	0.55103	268.58
220	1	22.309	66.314	111.14	3.2281	0.31681	0.54731	275.34
230	1	21.272	69.587	116.6	3.2523	0.3162	0.54423	281.89
240	1	20.332	72.843	122.03	3.2754	0.3157	0.54166	288.26
250	1	19.475	76.084	127.43	3.2975	0.31529	0.53948	294.47
260	1	18.69	79.313	132.82	3.3186	0.31495	0.53762	300.52
270	1	17.968	82.532	138.18	3.3389	0.31466	0.53602	306.44
280	1	17.302	85.741	143.54	3.3583	0.31441	0.53463	312.22
290	1	16.685	88.943	148.88	3.3771	0.3142	0.53342	317.88
300	1	16.111	92.138	154.21	3.3952	0.31402	0.53235	323.44
400	1	12.014	123.85	207.09	3.5473	0.31308	0.52632	374.06
500	1	9.5926	155.34	259.58	3.6645	0.31274	0.52388	418.25
600	1	7.9885	186.72	311.9	3.7599	0.31258	0.52266	458.06
700	1	6.8458	218.06	364.13	3.8404	0.31249	0.52197	494.61
800	1	5.99	249.36	416.31	3.91	0.31243	0.52153	528.59
900	1	5.3248	280.64	468.44	3.9715	0.3124	0.52125	560.5
1000	1	4.7929	311.91	520.56	4.0264	0.31237	0.52105	590.67

(continued)

Table 11.6 (continued)

Temperature (K)	Pressure (MPa)	Density (kg/m^3)	Internal energy (kJ/kg)	Enthalpy (kJ/kg)	Entropy (kJ/kg K)	C_v (kJ/kg K)	C_p (kJ/kg K)	Sound speed (m/s)
2.0 MPa								
84.29	2	1419.1	−121.43	−120.02	1.3301	0.55001	1.1092	868.34
90	2	1384.4	−115.13	−113.68	1.4029	0.52901	1.1126	829.49
100	2	1320.5	−103.96	−102.44	1.5213	0.49953	1.1406	758.33
110	2	1250.2	−92.376	−90.776	1.6324	0.47599	1.1984	681.29
120	2	1169.4	−80.014	−78.304	1.7409	0.45791	1.3085	594.4
129.74	2	1070.9	−66.431	−64.564	1.8509	0.44925	1.5548	490.64
129.74	2	102.11	25.837	45.424	2.6986	0.42622	1.1595	184.89
130	2	101.54	26.033	45.73	2.701	0.42403	1.1454	185.42
140	2	85.84	32.187	55.486	2.7734	0.37635	0.86177	201.78
150	2	75.879	37.13	63.488	2.8287	0.35566	0.75147	214.39
160	2	68.6	41.523	70.678	2.8751	0.34397	0.69171	225.19
170	2	62.905	45.601	77.395	2.9158	0.33653	0.65424	234.89
180	2	58.263	49.475	83.802	2.9525	0.33147	0.62868	243.8
190	2	54.37	53.207	89.992	2.9859	0.32784	0.61022	252.13
200	2	51.039	56.836	96.022	3.0169	0.32515	0.59636	259.99
210	2	48.144	60.387	101.93	3.0457	0.3231	0.58561	267.48
220	2	45.596	63.877	107.74	3.0727	0.3215	0.57708	274.65
230	2	43.331	67.32	113.48	3.0982	0.32023	0.57017	281.54
240	2	41.3	70.723	119.15	3.1224	0.31921	0.56449	288.2
250	2	39.467	74.094	124.77	3.1453	0.31837	0.55975	294.66
260	2	37.801	77.438	130.35	3.1672	0.31767	0.55575	300.92
270	2	36.28	80.759	135.89	3.1881	0.31709	0.55234	307.01
280	2	34.884	84.061	141.39	3.2081	0.3166	0.54941	312.96
290	2	33.597	87.347	146.88	3.2274	0.31618	0.54687	318.76
300	2	32.407	90.618	152.33	3.2459	0.31582	0.54465	324.43
400	2	24.028	122.83	206.07	3.4005	0.31395	0.53229	375.69
500	2	19.15	154.59	259.02	3.5187	0.31328	0.52739	420.08
600	2	15.938	186.14	311.63	3.6146	0.31296	0.52496	459.93
700	2	13.656	217.59	364.05	3.6954	0.31278	0.52358	496.46
800	2	11.948	248.98	416.36	3.7653	0.31267	0.52272	530.4
900	2	10.622	280.32	468.6	3.8268	0.31259	0.52215	562.25
1000	2	9.5622	311.64	520.8	3.8818	0.31254	0.52176	592.36
5.0 MPa								
85.037	5	1422.6	−121.34	−117.83	1.3311	0.5507	1.0996	877.21
90	5	1393.4	−115.96	−112.37	1.3934	0.53263	1.1003	844.78
100	5	1332	−105.04	−101.29	1.5102	0.50306	1.1202	777.34
110	5	1265.6	−93.844	−89.893	1.6188	0.47923	1.163	705.93
120	5	1191.6	−82.118	−77.922	1.7229	0.46001	1.2384	628.46
130	5	1104.8	−69.438	−64.912	1.827	0.44645	1.3801	541
140	5	991.96	−54.728	−49.688	1.9396	0.44418	1.728	433.54
150	5	765.37	−31.201	−24.668	2.1113	0.52622	5.1511	248.19
160	5	243.26	21.261	41.815	2.5466	0.43817	1.7602	206.39
170	5	197.63	30.372	55.672	2.6308	0.39055	1.149	223.58

(continued)

Table 11.6 (continued)

Temperature (K)	Pressure (MPa)	Density (kg/m³)	Internal energy (kJ/kg)	Enthalpy (kJ/kg)	Entropy (kJ/kg K)	C_v (kJ/kg K)	C_p (kJ/kg K)	Sound speed (m/s)
180	5	172.07	36.916	65.974	2.6898	0.3685	0.93849	236.49
190	5	154.47	42.398	74.766	2.7373	0.35536	0.82973	247.45
200	5	141.19	47.292	82.706	2.7781	0.34664	0.76303	257.2
210	5	130.6	51.813	90.098	2.8141	0.34046	0.71797	266.11
220	5	121.36	56.078	97.107	2.8468	0.33589	0.68557	274.4
230	5	114.47	60.156	103.84	2.8767	0.33239	0.66122	282.2
240	5	108.09	64.094	110.35	2.9044	0.32966	0.6423	289.6
250	5	102.52	67.922	116.69	2.9303	0.32746	0.62722	296.66
260	5	97.577	71.662	122.9	2.9547	0.32568	0.61496	303.44
270	5	93.161	75.33	129	2.9777	0.32421	0.60481	309.98
280	5	89.182	78.94	135	2.9995	0.32298	0.59629	316.29
290	5	85.571	82.5	140.93	3.0203	0.32194	0.58907	322.42
300	5	82.275	86.018	146.79	3.0402	0.32105	0.58287	328.37
400	5	60.011	119.81	203.12	3.2024	0.31649	0.55001	381.06
500	5	47.592	152.37	257.43	3.3237	0.31487	0.53769	425.81
600	5	39.543	184.42	310.86	3.4211	0.31409	0.53168	465.68
700	5	33.866	216.2	363.85	3.5028	0.31365	0.52828	502.09
800	5	29.634	247.84	416.56	3.5732	0.31337	0.52618	535.87
900	5	26.352	279.37	469.11	3.6351	0.31318	0.52479	567.54
1000	5	23.73	310.83	521.54	3.6903	0.31304	0.52383	597.47
10.0 MPa								
86.275	10	1428.2	−121.17	−114.17	1.3328	0.55191	1.085	891.32
90	10	1407.2	−117.24	−110.13	1.3787	0.53857	1.0834	868.4
100	10	1349.4	−106.68	−99.265	1.4931	0.50893	1.0932	805.91
110	10	1288.2	−95.979	−88.216	1.5984	0.48494	1.1193	741.46
120	10	1222.1	−85.003	−76.82	1.6976	0.465	1.1632	674.25
130	10	1149	−73.569	−64.866	1.7932	0.44893	1.2331	603.33
140	10	1065.5	−61.392	−52.007	1.8885	0.43737	1.3491	527.43
150	10	964.88	−47.94	−37.576	1.9879	0.43203	1.5594	445.1
160	10	833.57	−32.151	−20.155	2.1002	0.43563	1.97	357.76
170	10	656.74	−12.791	2.4358	2.237	0.44268	2.4969	284.47
180	10	491.85	5.816	26.147	2.3727	0.42542	2.0914	259.27
190	10	395.21	18.695	43.998	2.4693	0.4003	1.5271	260.68
200	10	337.74	27.944	57.553	2.5389	0.3815	1.2152	267.7
210	10	299.16	35.313	68.74	2.5935	0.36815	1.0373	275.72
220	10	270.9	41.601	78.516	2.639	0.35848	0.9257	283.74
230	10	248.97	47.206	87.372	2.6784	0.35126	0.85014	291.51
240	10	231.24	52.345	95.589	2.7134	0.34573	0.79597	298.98
250	10	216.5	57.148	103.34	2.745	0.34138	0.75542	306.16
260	10	203.96	61.7	110.73	2.774	0.3379	0.72404	313.06
270	10	193.12	66.057	117.84	2.8008	0.33505	0.69911	319.72
280	10	183.6	70.26	124.73	2.8259	0.3327	0.67887	326.15
290	10	175.16	74.338	131.43	2.8494	0.33072	0.66216	332.37
300	10	167.6	78.313	137.98	2.8716	0.32905	0.64815	338.42

(continued)

Table 11.6 (continued)

Temperature (K)	Pressure (MPa)	Density (kg/m³)	Internal energy (kJ/kg)	Enthalpy (kJ/kg)	Entropy (kJ/kg K)	C_v (kJ/kg K)	C_p (kJ/kg K)	Sound speed (m/s)
400	10	119.43	114.86	198.59	3.0465	0.32049	0.57843	391.51
500	10	94.094	148.76	255.04	3.1725	0.31743	0.55395	436.14
600	10	78.026	181.63	309.79	3.2724	0.31594	0.54225	475.69
700	10	66.802	213.96	363.66	3.3554	0.31508	0.5357	511.73
800	10	58.472	245.99	417.01	3.4267	0.31454	0.53165	545.13
900	10	52.026	277.82	470.03	3.4891	0.31415	0.52898	576.44
1000	10	46.88	309.52	522.83	3.5448	0.31388	0.52712	606.03
50.0 MPa								
95.804	50	1466.5	−119.16	−85.067	1.3491	0.56038	1.0063	987.92
100	50	1448.5	−115.38	−80.859	1.3921	0.54847	0.99945	969.74
110	50	1405.8	−106.49	−70.924	1.4868	0.52417	0.9886	927.77
120	50	1363	−97.758	−61.074	1.5725	0.50375	0.98198	887.55
130	50	1320.1	−89.152	−51.277	1.6509	0.48605	0.97749	849.02
140	50	1277.3	−80.666	−41.52	1.7232	0.47053	0.9742	812.28
150	50	1234.3	−72.299	−31.791	1.7903	0.45684	0.97158	777.44
160	50	1191.5	−64.051	−22.087	1.8529	0.44474	0.96923	744.66
170	50	1148.9	−55.929	−12.407	1.9116	0.43402	0.96676	714.08
180	50	1106.5	−47.94	−2.754	1.9668	0.42449	0.96374	685.82
190	50	1064.8	−40.095	6.8643	2.0188	0.41599	0.9597	659.98
200	50	1023.7	−32.407	16.435	2.0679	0.40839	0.95418	636.63
210	50	983.63	−24.89	25.942	2.1143	0.40159	0.94684	615.8
220	50	944.74	−17.559	35.365	2.1581	0.39548	0.93753	597.47
230	50	907.26	−10.425	44.686	2.1996	0.38999	0.92634	581.56
240	50	871.36	−3.4948	53.887	2.2387	0.38504	0.91352	567.93
250	50	837.16	3.2265	62.952	2.2757	0.38055	0.89942	556.4
260	50	804.73	9.7394	71.872	2.3107	0.37647	0.88438	546.77
270	50	774.1	16.047	80.638	2.3438	0.37275	0.86875	538.85
280	50	745.27	22.156	89.246	2.3751	0.36933	0.85284	532.42
290	50	718.18	28.074	97.695	2.4048	0.36619	0.8369	527.3
300	50	692.78	33.812	105.98	2.4329	0.3633	0.82119	523.31
400	50	511.19	83.58	181.39	2.6506	0.34414	0.69869	519.74
500	50	408.27	125.22	247.68	2.7988	0.33469	0.63409	541.31
600	50	342.22	163.06	309.16	2.911	0.32929	0.59866	568.28
700	50	295.84	198.87	367.88	3.0016	0.32585	0.57733	596.07
800	50	261.22	233.47	424.87	3.0777	0.32349	0.56351	623.42
900	50	234.26	267.28	480.72	3.1435	0.32179	0.55404	649.96
1000	50	212.57	300.56	535.77	3.2015	0.3205	0.54728	675.63
100.0 MPa								
106.96	100	1504.6	−115.72	−49.256	1.3704	0.56824	0.95295	1084.1
110	100	1494.3	−113.29	−46.364	1.3971	0.56101	0.94752	1074.3
120	100	1460.7	−105.43	−36.967	1.4789	0.53995	0.93255	1043.4
130	100	1427.7	−97.747	−27.705	1.553	0.52192	0.92014	1014.4
140	100	1395.3	−90.229	−18.559	1.6208	0.50609	0.90911	987
150	100	1363.4	−82.865	−9.5198	1.6832	0.49202	0.89893	961.21

(continued)

Table 11.6 (continued)

Temperature (K)	Pressure (MPa)	Density (kg/m^3)	Internal energy (kJ/kg)	Enthalpy (kJ/kg)	Entropy (kJ/kg K)	C_v (kJ/kg K)	C_p (kJ/kg K)	Sound speed (m/s)
160	100	1332.1	−75.647	-0.57889	1.7409	0.47944	0.88934	936.96
170	100	1301.4	−68.57	8.2684	1.7945	0.46814	0.88018	914.2
180	100	1271.4	−61.63	17.026	1.8446	0.45798	0.87138	892.89
190	100	1241.9	−54.824	25.697	1.8915	0.44881	0.86285	873
200	100	1213.1	−48.148	34.284	1.9355	0.44052	0.85453	854.48
210	100	1185	−41.598	42.788	1.977	0.43301	0.84638	837.28
220	100	1157.6	−35.173	51.212	2.0162	0.42618	0.83834	821.35
230	100	1130.9	−28.869	59.555	2.0533	0.41997	0.83039	806.65
240	100	1104.9	−22.682	67.82	2.0884	0.41431	0.8225	793.12
250	100	1079.7	−16.611	76.005	2.1219	0.40913	0.81465	780.71
260	100	1055.2	−10.652	84.113	2.1537	0.40438	0.80684	769.35
270	100	1031.5	−4.8028	92.142	2.184	0.40001	0.79907	759
280	100	1008.5	0.94075	100.09	2.2129	0.39599	0.79135	749.6
290	100	986.31	6.5814	107.97	2.2405	0.39228	0.7837	741.08
300	100	964.82	12.123	115.77	2.267	0.38885	0.77613	733.39
400	100	787.37	62.839	189.84	2.4805	0.36495	0.7081	690.42
500	100	663.41	107.26	258	2.6328	0.35165	0.65805	683.08
600	100	574.15	147.81	321.98	2.7496	0.34335	0.62368	690.94
700	100	507.15	185.91	383.09	2.8438	0.33774	0.6	705.52
800	100	454.97	222.41	442.21	2.9228	0.33372	0.58324	723.21
900	100	413.09	257.81	499.89	2.9907	0.33071	0.57101	742.34
1000	100	378.65	292.42	556.51	3.0504	0.32838	0.56185	762.09

Reference

1. Ch. Tegeler, R. Span, W. Wagner, A new equation of state for argon covering the fluid region for temperatures from the melting line to 700 K at pressures up to 1000 MPa. J. Phys. Chem. Ref. Data **28**(3), 779–850 (1999)

Chapter 12
Oxygen

12.1 Formulation for Oxygen

The formulation for oxygen is one of the oldest within this text, and was published by Schmidt and Wagner [1]. Because it was developed before 1990, it remains one of the few equations of state with temperatures reported on the ITPS-68 temperature scale. As such, the 0.05% uncertainty in vapor pressure with temperatures on the IPTS-68 scale increases to 0.15% at low temperatures near the triple point and to 0.1% at higher temperatures. To obtain the 0.05% uncertainty in vapor pressure, temperatures in the tables given here must be converted to the ITS-90 temperature scale.

Oxygen is currently in the process of being re-fitted, although this work will take up to an additional year to complete and is not ready for use in this book. Due to safety concerns handling oxygen at high temperatures, experimental measurements are available only to 323 K. One data set is available with densities up to 673 K, but these measurements have uncertainties up to 10% and are not useful in the development of the new equation of state. Instead, calculations from statistical thermodynamics are currently being made to aid in the development of the equation at high temperatures and pressures. The upper temperature listed in Table 12.1 is 2000 K, which was verified by use of corresponding states with comparisons to the data of nitrogen.

The uncertainty statement in REFPROP is as follows.

The uncertainties of the equation of state are 0.15% in vapor pressure, 0.1% in density, 2% in heat capacity, and 1% in the speed of sound, except in the critical region (Tables 12.1, 12.2, 12.3 and 12.4).

J.W. Leachman et al., *Thermodynamic Properties of Cryogenic Fluids*,
International Cryogenics Monograph Series, DOI 10.1007/978-3-319-57835-4_12

Table 12.1 Fixed point properties and correlation limits for oxygen

	Temperature (K)	Pressure (kPa)	Density (mol/dm^3)
Critical point	154.581	5043.0	13.63
Triple point	54.361	0.14628	40.816
Upper limit	2000	82,000	43.348
Lower limit	54.361		

Table 12.2 Fluid constants and normal boiling point for oxygen

Molar mass	31.9988 g/mol
Ideal-gas constant	8.31434 J/(mol K)
Pitzer's acentric factor	0.0222
Normal boiling point	90.1878 K

Table 12.3 Fundamental equation of state coefficients and exponents for oxygen

i	N_i	t_i	d_i	p_i
1	0.39837687490	0	1	
2	−1.8461574540	1.5	1	
3	0.41834731970	2.5	1	
4	0.023706207110	−0.5	2	
5	0.097717305730	1.5	2	
6	0.030178912940	2	2	
7	0.022733532120	0	3	
8	0.013572540860	1	3	
9	−0.040526989430	2.5	3	
10	0.00054546285150	0	6	
11	0.00051131822770	2	7	
12	0.00000029534668830	5	7	
13	−0.000086876450720	2	8	
14	−0.21270825890	5	1	2
15	0.087359419580	6	1	2
16	0.1275509190	3.5	2	2
17	−0.090677010640	5.5	2	2
18	−0.035400842060	3	3	2
19	−0.036232780590	7	3	2
20	0.01327699290	6	5	2
21	−0.00032541118650	8.5	6	2
22	−0.0083135829320	4	7	2
23	0.0021245705590	6.5	8	2
24	−0.00083252062320	5.5	10	2

(continued)

Table 12.3 (continued)

i	N_i	t_i	d_i	p_i
25	−0.000026261732760	22	2	4
26	0.0025995814820	11	3	4
27	0.0099846496630	18	3	4
28	0.0021999231530	11	4	4
29	−0.025913504860	23	4	4
30	−0.12596308480	17	5	4
31	0.14783556370	18	5	4
32	−0.010112510780	23	5	4

Table 12.4 Ancillary function coefficients and exponents for oxygen

	α^0	
i	a_i	k_i
0	3.518087320	
1	−14.7168366665	
2	−0.0110839854	
3	1.02323928	2246.3244
4	0.734357918	11259.9763
5	0.00337183363	1201.26209
6	−0.0170864084	69.0089445
7	0.0463751562	5328.05445

$n = 7$ (there are no polynomial terms in the first summation)

	Vapor pressure		Saturated liquid density		Saturated vapor density	
i	N_i	k_i	N_i	k_i	N_i	k_i
1	−6.0595	1	1.6622	0.345	−2.2695	0.3785
2	1.305	1.5	0.76846	0.74	−4.6578	1.07
3	−0.54178	2.2	−0.10041	1.2	−9.948	2.7
4	−1.941	4.8	0.2048	2.6	−22.845	5.5
5	0.35514	6.2	0.011551	7.2	−45.19	10
6					−25.101	20

12.2 Property Plots for Oxygen

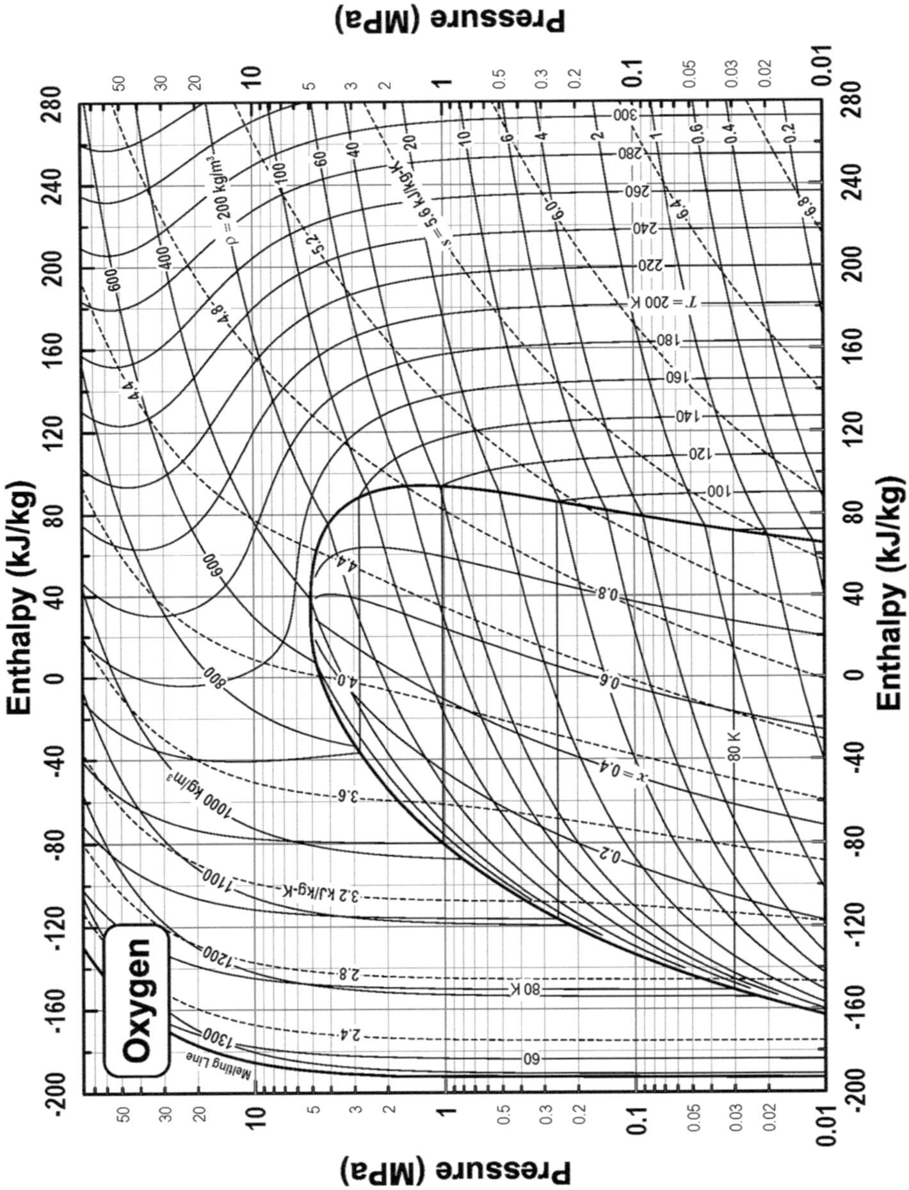
Oxygen
Pressure (MPa)
Enthalpy (kJ/kg)
Melting Line

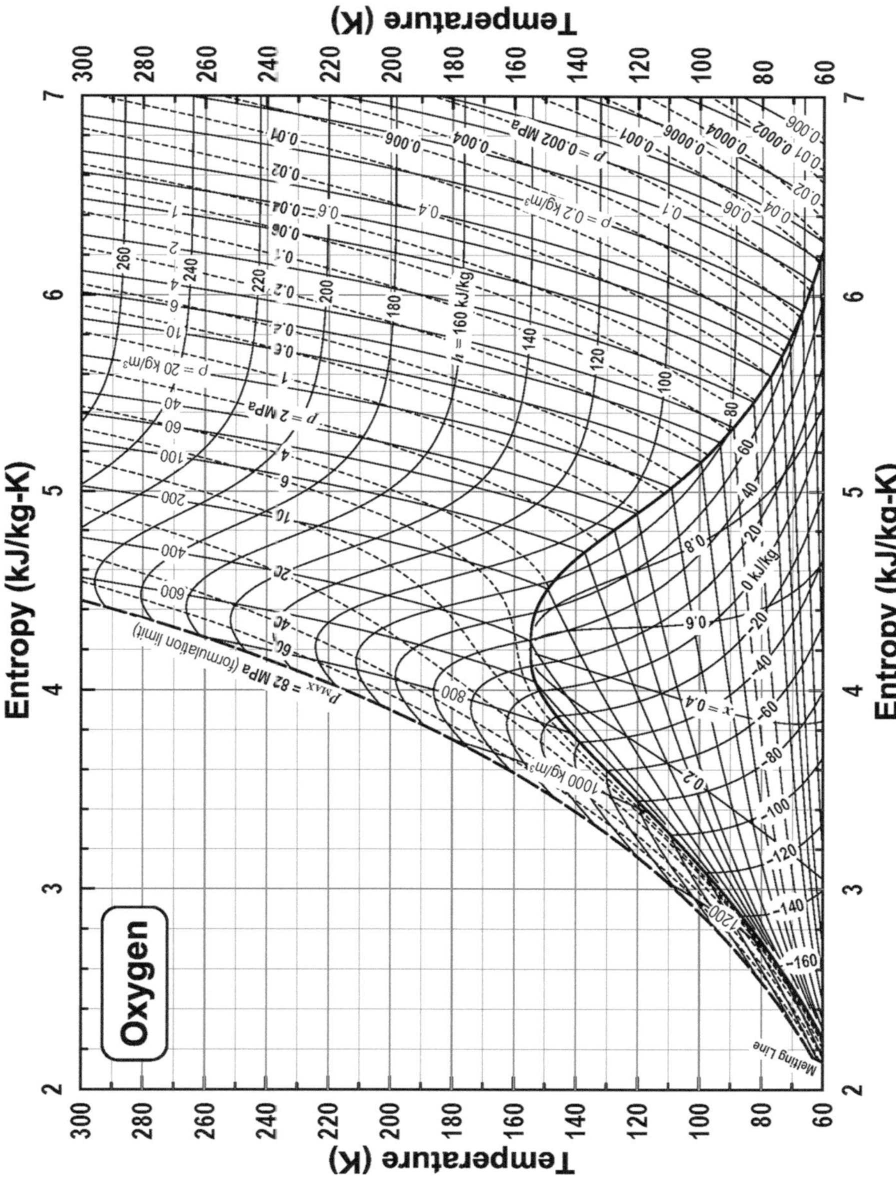
Oxygen
Entropy (kJ/kg-K)
Temperature (K)
p_{MAX} = 82 MPa (formulation limit)
p = 2 MPa
p = 0.002 MPa
ρ = 20 kg/m³
ρ = 0.2 kg/m³
1000 kg/m³
h = 160 kJ/kg
0 kJ/kg
x = 0.4
Melting Line

12.3 Thermodynamic Data Tables for Oxygen

See Tables 12.5 and 12.6.

Table 12.5 Thermodynamic properties of saturated oxygen

Temperature (K)	Pressure (MPa)	Liquid density (kg/m^3)	Vapor density (kg/m^3)	Liquid enthalpy (kJ/kg)	Vapor enthalpy (kJ/kg)	Liquid entropy (kJ/kg K)	Vapor entropy (kJ/kg K)	Liquid C_v (kJ/kg K)	Vapor C_v (kJ/kg K)	Liquid sound speed (m/s)	Vapor sound speed (m/s)
54.361	0.000146	1306.1	0.010358	−193.61	49.107	2.0921	6.5571	1.1954	0.66381	1123.4	140.32
58	0.000428	1290.7	0.028382	−187.53	52.39	2.2004	6.337	1.1149	0.67492	1131.8	144.71
62	0.001187	1273.1	0.073791	−180.84	55.98	2.312	6.1317	1.0688	0.68846	1119	149.31
66	0.002882	1255.2	0.16844	−174.13	59.55	2.4168	5.9575	1.0395	0.69955	1095.1	153.75
70	0.006262	1237	0.34573	−167.42	63.092	2.5156	5.8086	1.0167	0.70519	1066.3	158.07
74	0.012414	1218.5	0.65008	−160.7	66.589	2.6088	5.6803	0.99678	0.70458	1035.4	162.29
78	0.022792	1199.9	1.1367	−153.98	70.017	2.6972	5.5689	0.97845	0.69896	1003.5	166.38
82	0.039226	1181	1.8709	−147.24	73.342	2.7812	5.4713	0.96124	0.69069	971.26	170.29
86	0.063906	1161.7	2.9268	−140.48	76.532	2.8614	5.3849	0.94496	0.68227	938.71	173.95
90.188	0.10133	1141.2	4.4671	−133.37	79.688	2.9419	5.3042	0.92887	0.67555	904.35	177.45
94	0.14836	1122	6.3426	−126.84	82.369	3.0123	5.2379	0.91506	0.67256	872.75	180.3
98	0.21397	1101.4	8.8928	−119.93	84.957	3.0836	5.1743	0.9014	0.67329	839.17	182.91
102	0.29941	1080.2	12.148	−112.94	87.285	3.1528	5.1158	0.88861	0.67809	805.06	185.09
106	0.40806	1058.2	16.23	−105.85	89.328	3.22	5.0613	0.87672	0.68673	770.3	186.84
110	0.5434	1035.5	21.281	−98.641	91.054	3.2855	5.01	0.86577	0.69881	734.76	188.14
114	0.70902	1011.7	27.464	−91.283	92.432	3.3498	4.9613	0.85582	0.71391	698.3	188.99
118	0.90859	986.81	34.978	−83.749	93.42	3.413	4.9145	0.84699	0.73166	660.76	189.38
122	1.1459	960.52	44.068	−76.002	93.97	3.4756	4.8688	0.83945	0.75188	621.94	189.32
126	1.4247	932.55	55.055	−67.994	94.015	3.5378	4.8235	0.83343	0.77461	581.62	188.78
130	1.7491	902.48	68.369	−59.662	93.466	3.6001	4.778	0.82929	0.80017	539.5	187.75
134	2.1232	869.7	84.628	−50.916	92.195	3.6631	4.7311	0.82763	0.82934	495.18	186.21
138	2.5516	833.28	104.77	−41.617	90.006	3.7278	4.6816	0.82948	0.8636	448.03	184.11
142	3.0394	791.6	130.38	−31.538	86.574	3.7955	4.6273	0.83708	0.90582	397.03	181.36
146	3.5925	741.55	164.53	−20.242	81.274	3.8689	4.5642	0.85614	0.96213	340.34	177.79
150	4.2186	675.48	214.94	−6.6709	72.562	3.9546	4.4828	0.9057	1.0492	273.8	172.83
154.5811	5.0428	436.14	436.14	32.423	32.423	4.2008	4.2008				

Table 12.6 Thermodynamic properties of oxygen

Temperature (K)	Pressure (MPa)	Density (kg/m³)	Internal energy (kJ/kg)	Enthalpy (kJ/kg)	Entropy (kJ/kg K)	C_v (kJ/kg K)	C_p (kJ/kg K)	Sound speed (m/s)
0.1 MPa								
54.371	0.1	1306.1	−193.61	−193.53	2.0922	1.1948	1.6728	1123.8
60	0.1	1282.1	−184.21	−184.13	2.2569	1.0887	1.6732	1127.7
70	0.1	1237.1	−167.44	−167.36	2.5153	1.0168	1.6779	1066.6
80	0.1	1190.6	−150.66	−150.57	2.7395	0.96986	1.6814	987.7
90.062	0.1	1141.8	−133.67	−133.58	2.9395	0.92934	1.699	905.38
90.062	0.1	4.4135	56.939	79.597	5.3065	0.6757	0.97053	177.35
100	0.1	3.9411	63.612	88.986	5.4054	0.65269	0.9352	188.37
110	0.1	3.5625	70.247	98.318	5.4944	0.65413	0.93155	198.11
120	0.1	3.2524	76.87	107.62	5.5753	0.65431	0.92799	207.35
130	0.1	2.9935	83.473	116.88	5.6494	0.65371	0.92461	216.16
140	0.1	2.7736	90.056	126.11	5.7178	0.653	0.92182	224.59
150	0.1	2.5845	96.624	135.32	5.7814	0.65237	0.91964	232.7
160	0.1	2.4199	103.18	144.5	5.8407	0.6519	0.91797	240.51
170	0.1	2.2752	109.73	153.68	5.8963	0.65155	0.9167	248.05
180	0.1	2.1471	116.26	162.84	5.9486	0.65134	0.91575	255.36
190	0.1	2.0328	122.8	171.99	5.9981	0.65124	0.91506	262.45
200	0.1	1.9301	129.33	181.14	6.0451	0.65125	0.91458	269.34
210	0.1	1.8374	135.86	190.29	6.0897	0.65137	0.91431	276.05
220	0.1	1.7532	142.39	199.43	6.1322	0.65161	0.91421	282.58
230	0.1	1.6764	148.92	208.57	6.1728	0.65198	0.91429	288.95
240	0.1	1.6062	155.45	217.71	6.2118	0.65248	0.91455	295.17
250	0.1	1.5416	161.99	226.86	6.2491	0.65313	0.91499	301.25
260	0.1	1.482	168.54	236.01	6.285	0.65393	0.9156	307.19
270	0.1	1.4269	175.09	245.17	6.3196	0.65488	0.9164	313.01
280	0.1	1.3757	181.65	254.34	6.3529	0.656	0.91737	318.69
290	0.1	1.3281	188.23	263.52	6.3851	0.65728	0.91853	324.27
300	0.1	1.2837	194.81	272.71	6.4163	0.65873	0.91987	329.72
400	0.1	0.96217	261.75	365.68	6.6836	0.68116	0.94165	379.05
500	0.1	0.76959	331.39	461.33	6.8969	0.71193	0.97215	421.27
600	0.1	0.64128	404.18	560.12	7.0769	0.74317	1.0033	458.87
700	0.1	0.54965	479.94	661.87	7.2337	0.77103	1.031	493.31
800	0.1	0.48095	558.26	766.18	7.373	0.79455	1.0545	525.38
900	0.1	0.42751	638.73	872.64	7.4983	0.81402	1.0739	555.59
1000	0.1	0.38476	720.96	980.86	7.6123	0.83013	1.09	584.26
0.2 MPa								
54.382	0.2	1306.2	−193.61	−193.45	2.0923	1.1942	1.6724	1124.1
60	0.2	1282.2	−184.22	−184.07	2.2566	1.0887	1.673	1128
70	0.2	1237.2	−167.46	−167.3	2.515	1.017	1.6777	1066.9
80	0.2	1190.8	−150.68	−150.52	2.7391	0.97006	1.681	988.08
90	0.2	1142.3	−133.81	−133.63	2.9379	0.92978	1.6983	906.38
97.236	0.2	1105.4	−121.44	−121.26	3.0702	0.90394	1.7243	845.63
97.236	0.2	8.3545	60.542	84.481	5.1861	0.67283	0.99114	182.44
100	0.2	8.0871	62.463	87.194	5.2136	0.66264	0.9736	185.71

(continued)

Table 12.6 (continued)

Temperature (K)	Pressure (MPa)	Density (kg/m³)	Internal energy (kJ/kg)	Enthalpy (kJ/kg)	Entropy (kJ/kg K)	C_v (kJ/kg K)	C_p (kJ/kg K)	Sound speed (m/s)
110	0.2	7.262	69.28	96.821	5.3054	0.65893	0.9563	196.12
120	0.2	6.6005	76.04	106.34	5.3882	0.65891	0.94781	205.73
130	0.2	6.0558	82.754	115.78	5.4638	0.65763	0.94042	214.84
140	0.2	5.5982	89.428	125.15	5.5332	0.65614	0.93448	223.51
150	0.2	5.2074	96.067	134.47	5.5975	0.65485	0.9299	231.81
160	0.2	4.8692	102.68	143.76	5.6574	0.65385	0.9264	239.77
170	0.2	4.5734	109.27	153.01	5.7135	0.65311	0.92373	247.45
180	0.2	4.3123	115.85	162.23	5.7662	0.65259	0.92169	254.86
190	0.2	4.08	122.42	171.44	5.816	0.65226	0.92015	262.04
200	0.2	3.8718	128.98	180.64	5.8632	0.6521	0.91899	269
210	0.2	3.6841	135.53	189.82	5.908	0.65208	0.91816	275.77
220	0.2	3.514	142.09	199	5.9507	0.65222	0.91761	282.36
230	0.2	3.3591	148.64	208.17	5.9915	0.65251	0.91732	288.78
240	0.2	3.2175	155.19	217.35	6.0305	0.65294	0.91726	295.04
250	0.2	3.0873	161.74	226.52	6.068	0.65353	0.91743	301.15
260	0.2	2.9674	168.3	235.7	6.104	0.65429	0.91781	307.12
270	0.2	2.8566	174.86	244.88	6.1386	0.65521	0.91841	312.96
280	0.2	2.7537	181.44	254.07	6.172	0.65629	0.91921	318.67
290	0.2	2.6581	188.02	263.26	6.2043	0.65755	0.92022	324.26
300	0.2	2.5689	194.62	272.47	6.2355	0.65897	0.92142	329.74
400	0.2	1.9244	261.61	365.54	6.5031	0.68128	0.94243	379.16
500	0.2	1.5389	331.3	461.26	6.7166	0.71201	0.97261	421.42
600	0.2	1.2822	404.11	560.09	6.8967	0.74322	1.0036	459.04
700	0.2	1.099	479.88	661.86	7.0535	0.77107	1.0312	493.48
800	0.2	0.96163	558.21	766.19	7.1928	0.79457	1.0547	525.56
900	0.2	0.85479	638.68	872.66	7.3182	0.81404	1.0741	555.77
1000	0.2	0.76933	720.93	980.89	7.4322	0.83015	1.0901	584.43
0.5 MPa								
54.417	0.5	1306.4	−193.59	−193.2	2.0927	1.1925	1.6711	1125.3
60	0.5	1282.6	−184.27	−183.88	2.2558	1.0888	1.6723	1128.8
70	0.5	1237.7	−167.53	−167.12	2.5141	1.0175	1.677	1067.8
80	0.5	1191.3	−150.76	−150.34	2.7381	0.97065	1.6801	989.22
90	0.5	1143	−133.91	−133.47	2.9368	0.93039	1.6968	907.83
100	0.5	1091.6	−116.79	−116.34	3.1173	0.89539	1.7354	823.74
108.81	0.5	1042.4	−101.29	−100.81	3.2661	0.86893	1.7964	745.46
108.81	0.5	19.663	65.146	90.574	5.025	0.69487	1.0865	187.8
110	0.5	19.373	66.053	91.862	5.0368	0.68849	1.0718	189.32
120	0.5	17.309	73.377	102.26	5.1274	0.67445	1.0196	200.58
130	0.5	15.704	80.49	112.33	5.2079	0.6703	0.9948	210.72
140	0.5	14.404	87.469	122.18	5.281	0.66615	0.9767	220.18
150	0.5	13.322	94.347	131.88	5.3479	0.66264	0.96333	229.1
160	0.5	12.404	101.15	141.46	5.4097	0.65992	0.95341	237.55
170	0.5	11.612	107.9	150.95	5.4673	0.65789	0.94597	245.62
180	0.5	10.921	114.6	160.38	5.5212	0.65642	0.94032	253.36

(continued)

Table 12.6 (continued)

Temperature (K)	Pressure (MPa)	Density (kg/m^3)	Internal energy (kJ/kg)	Enthalpy (kJ/kg)	Entropy (kJ/kg K)	C_v (kJ/kg K)	C_p (kJ/kg K)	Sound speed (m/s)
190	0.5	10.311	121.27	169.77	5.5719	0.65537	0.93597	260.8
200	0.5	9.7682	127.92	179.11	5.6198	0.65466	0.93262	267.99
210	0.5	9.2813	134.55	188.42	5.6652	0.65424	0.93002	274.95
220	0.5	8.8431	141.17	197.71	5.7085	0.65405	0.92804	281.7
230	0.5	8.4451	147.78	206.98	5.7497	0.65408	0.92656	288.25
240	0.5	8.0824	154.38	216.24	5.7891	0.65432	0.92552	294.63
250	0.5	7.7501	160.98	225.49	5.8269	0.65475	0.92485	300.85
260	0.5	7.4447	167.58	234.74	5.8631	0.65537	0.92452	306.91
270	0.5	7.1629	174.18	243.99	5.898	0.65617	0.9245	312.83
280	0.5	6.902	180.79	253.23	5.9316	0.65717	0.92478	318.61
290	0.5	6.6597	187.4	262.48	5.9641	0.65834	0.92532	324.26
300	0.5	6.4341	194.03	271.74	5.9955	0.6597	0.92611	329.79
400	0.5	4.8111	261.22	365.14	6.2641	0.68165	0.94478	379.52
500	0.5	3.8451	331.01	461.04	6.4779	0.71223	0.974	421.89
600	0.5	3.2031	403.88	559.98	6.6582	0.74337	1.0045	459.55
700	0.5	2.7453	479.7	661.83	6.8152	0.77117	1.0319	494
800	0.5	2.4021	558.06	766.21	6.9545	0.79465	1.0551	526.08
900	0.5	2.1353	638.56	872.72	7.08	0.8141	1.0744	556.28
1000	0.5	1.9219	720.82	980.98	7.194	0.8302	1.0904	584.93
1.0 MPa								
54.474	1	1306.7	−193.56	−192.79	2.0932	1.1895	1.669	1127.2
60	1	1283.2	−184.35	−183.57	2.2544	1.089	1.6712	1130.2
70	1	1238.4	−167.63	−166.82	2.5126	1.0184	1.676	1069.3
80	1	1192.2	−150.9	−150.06	2.7365	0.97164	1.6785	991.11
90	1	1144.1	−134.08	−133.21	2.9349	0.9314	1.6943	910.23
100	1	1093	−117.02	−116.1	3.1151	0.89639	1.7313	826.85
110	1	1037.2	−99.441	−98.477	3.283	0.8666	1.8001	738.58
119.62	1	976.34	−81.661	−80.637	3.4384	0.84377	1.9212	645.19
119.62	1	38.456	67.696	93.699	4.8958	0.73956	1.2677	189.41
120	1	38.246	68.031	94.178	4.8998	0.73552	1.2568	190
130	1	33.715	76.256	105.92	4.9939	0.69642	1.121	203.1
140	1	30.39	83.919	116.82	5.0747	0.68507	1.0655	214.24
150	1	27.779	91.292	127.29	5.1469	0.67691	1.0299	224.36
160	1	25.648	98.469	137.46	5.2126	0.6708	1.0052	233.74
170	1	23.861	105.51	147.42	5.273	0.66632	0.98742	242.53
180	1	22.333	112.45	157.22	5.329	0.66307	0.97429	250.84
190	1	21.007	119.31	166.91	5.3814	0.66072	0.96438	258.76
200	1	19.842	126.12	176.52	5.4307	0.65904	0.95675	266.33
210	1	18.808	132.89	186.05	5.4772	0.65789	0.95082	273.61
220	1	17.883	139.62	195.54	5.5213	0.65714	0.94617	280.63
230	1	17.05	146.33	204.98	5.5633	0.65673	0.94252	287.42
240	1	16.295	153.02	214.39	5.6033	0.65662	0.93969	294
250	1	15.607	159.7	223.78	5.6417	0.65677	0.93753	300.38
260	1	14.977	166.37	233.14	5.6784	0.65716	0.93594	306.59

(continued)

Table 12.6 (continued)

Temperature (K)	Pressure (MPa)	Density (kg/m^3)	Internal energy (kJ/kg)	Enthalpy (kJ/kg)	Entropy (kJ/kg K)	C_v (kJ/kg K)	C_p (kJ/kg K)	Sound speed (m/s)
270	1	14.397	173.04	242.5	5.7137	0.65778	0.93484	312.64
280	1	13.863	179.7	251.84	5.7477	0.65862	0.93418	318.53
290	1	13.367	186.37	261.18	5.7805	0.65966	0.93392	324.29
300	1	12.907	193.05	270.52	5.8121	0.66091	0.934	329.9
400	1	9.6226	260.56	364.48	6.0823	0.68226	0.94869	380.13
500	1	7.683	330.52	460.68	6.2969	0.7126	0.9763	422.68
600	1	6.3982	403.51	559.8	6.4775	0.74362	1.0059	460.4
700	1	5.483	479.4	661.78	6.6346	0.77135	1.0329	494.87
800	1	4.7976	557.81	766.25	6.7741	0.79478	1.0559	526.94
900	1	4.2649	638.35	872.82	6.8996	0.8142	1.075	557.14
1000	1	3.8388	720.63	981.13	7.0137	0.83028	1.0908	585.77
2.0 MPa								
54.588	2	1307.4	−193.49	−191.96	2.0943	1.1839	1.6651	1131
60	2	1284.4	−184.51	−182.95	2.2518	1.0894	1.669	1133
70	2	1239.8	−167.84	−166.22	2.5096	1.0201	1.6739	1072.4
80	2	1193.9	−151.16	−149.48	2.7332	0.9736	1.6754	994.85
90	2	1146.3	−134.42	−132.67	2.9312	0.93341	1.6894	914.97
100	2	1095.8	−117.45	−115.63	3.1107	0.89838	1.7233	832.95
110	2	1041	−100.03	−98.109	3.2776	0.86845	1.7862	746.7
120	2	979.15	−81.781	−79.738	3.4374	0.84432	1.8999	652.74
130	2	904.69	−61.936	−59.726	3.5975	0.82916	2.1358	543.94
132.74	2	880.34	−55.988	−53.716	3.6432	0.82783	2.2521	509.38
132.74	2	79.153	67.415	92.683	4.7461	0.81971	1.7407	186.75
140	2	69.863	75.176	103.8	4.8277	0.74234	1.3958	200.07
150	2	61.402	84.245	116.82	4.9176	0.71225	1.2299	213.92
160	2	55.343	92.515	128.65	4.994	0.69608	1.1448	225.69
170	2	50.654	100.33	139.82	5.0617	0.68516	1.0917	236.2
180	2	46.858	107.86	150.54	5.123	0.6775	1.0557	245.82
190	2	43.689	115.19	160.97	5.1794	0.67207	1.03	254.75
200	2	40.987	122.37	171.17	5.2317	0.66819	1.011	263.14
210	2	38.643	129.45	181.2	5.2807	0.66541	0.99661	271.09
220	2	36.584	136.44	191.11	5.3268	0.66344	0.98542	278.66
230	2	34.756	143.37	200.92	5.3704	0.6621	0.97661	285.91
240	2	33.12	150.26	210.65	5.4118	0.66126	0.96963	292.88
250	2	31.643	157.11	220.31	5.4512	0.66083	0.96407	299.6
260	2	30.303	163.93	229.93	5.489	0.66075	0.95965	306.1
270	2	29.079	170.73	239.51	5.5251	0.66099	0.95618	312.4
280	2	27.957	177.52	249.06	5.5598	0.66151	0.9535	318.51
290	2	26.923	184.3	258.58	5.5932	0.66229	0.9515	324.46
300	2	25.967	191.07	268.09	5.6255	0.6633	0.95008	330.25
400	2	19.244	259.25	363.17	5.899	0.68346	0.9565	381.41
500	2	15.337	329.56	459.97	6.1149	0.71334	0.98087	424.29
600	2	12.764	402.76	559.46	6.2962	0.74411	1.0089	462.13
700	2	10.936	478.79	661.68	6.4537	0.7717	1.035	496.62

(continued)

Table 12.6 (continued)

Temperature (K)	Pressure (MPa)	Density (kg/m^3)	Internal energy (kJ/kg)	Enthalpy (kJ/kg)	Entropy (kJ/kg K)	C_v (kJ/kg K)	C_p (kJ/kg K)	Sound speed (m/s)
800	2	9.5683	557.31	766.33	6.5934	0.79504	1.0574	528.69
900	2	8.5069	637.92	873.03	6.719	0.8144	1.0761	558.86
1000	2	7.6579	720.27	981.44	6.8333	0.83043	1.0917	587.46
5.0 MPa								
54.931	5	1309.2	−193.31	−189.49	2.0976	1.1688	1.6543	1141.8
60	5	1287.9	−184.97	−181.09	2.2439	1.0905	1.6627	1141.2
70	5	1244.1	−168.44	−164.42	2.5009	1.0251	1.6681	1081.3
80	5	1199.1	−151.92	−147.75	2.7235	0.97935	1.6669	1005.8
90	5	1152.6	−135.38	−131.05	2.9202	0.9393	1.676	928.7
100	5	1103.9	−118.7	−114.17	3.098	0.90425	1.7019	850.39
110	5	1051.6	−101.69	−96.934	3.2622	0.87401	1.7501	769.42
120	5	993.94	−84.088	−79.057	3.4177	0.84878	1.8325	683.68
130	5	927.76	−65.46	−60.071	3.5696	0.82982	1.9803	589.79
140	5	845.54	−44.826	−38.912	3.7263	0.82178	2.2996	480.62
150	5	717.26	−18.391	−11.42	3.9155	0.85378	3.6164	330.93
154.36	5	512.33	11.055	20.814	4.1263	1.215	111.62	163.89
154.36	5	357.09	32.169	46.171	4.2905	1.3096	132.86	158.85
160	5	207.36	63.61	87.722	4.5569	0.84959	3.0446	196.19
170	5	162.68	79.602	110.34	4.6943	0.76562	1.828	216.92
180	5	140.16	91.046	126.72	4.788	0.732	1.4958	231.86
190	5	125.18	100.85	140.79	4.8642	0.7119	1.3343	244.38
200	5	114.07	109.78	153.62	4.93	0.69867	1.2384	255.39
210	5	105.3	118.18	165.67	4.9888	0.68956	1.1751	265.34
220	5	98.123	126.23	177.18	5.0424	0.68314	1.1305	274.51
230	5	92.076	134.01	188.31	5.0918	0.67855	1.0976	283.05
240	5	86.879	141.61	199.16	5.138	0.67526	1.0726	291.09
250	5	82.344	149.06	209.78	5.1814	0.67295	1.053	298.71
260	5	78.338	156.41	220.23	5.2224	0.67139	1.0375	305.97
270	5	74.752	163.67	230.55	5.2613	0.67044	1.0251	312.92
280	5	71.545	170.86	240.74	5.2984	0.66998	1.015	319.6
290	5	68.629	178	250.85	5.3338	0.66994	1.0068	326.03
300	5	65.969	185.09	260.88	5.3679	0.67028	1.0001	332.25
400	5	48.063	255.33	359.36	5.6513	0.68693	0.97968	385.72
500	5	38.105	326.71	457.92	5.8712	0.71546	0.99426	429.36
600	5	31.658	400.55	558.49	6.0545	0.74554	1.0176	467.44
700	5	27.112	477.01	661.43	6.2131	0.77271	1.041	501.98
800	5	23.724	555.84	766.59	6.3535	0.79579	1.0617	534
900	5	21.097	636.67	873.67	6.4796	0.81497	1.0794	564.07
1000	5	18.998	719.19	982.38	6.5941	0.83088	1.0943	592.55
10.0 MPa								
55.497	10	1312.3	−193.02	−185.4	2.1027	1.1487	1.6395	1158.4
70	10	1250.9	−169.39	−161.4	2.4868	1.0331	1.6594	1095.8
80	10	1207.3	−153.11	−144.83	2.708	0.98849	1.6543	1023.3
90	10	1162.6	−136.88	−128.28	2.9029	0.94867	1.6569	950.23

(continued)

Table 12.6 (continued)

Temperature (K)	Pressure (MPa)	Density (kg/m^3)	Internal energy (kJ/kg)	Enthalpy (kJ/kg)	Entropy (kJ/kg K)	C_v (kJ/kg K)	C_p (kJ/kg K)	Sound speed (m/s)
100	10	1116.3	−120.61	−111.65	3.0782	0.91364	1.6725	877.07
110	10	1067.4	−104.15	−94.782	3.2389	0.88318	1.7035	802.97
120	10	1015	−87.363	−77.511	3.3892	0.85705	1.7546	726.96
130	10	957.54	−70.036	−59.592	3.5325	0.83541	1.8353	647.91
140	10	892.47	−51.845	−40.641	3.6729	0.81912	1.9663	564.48
150	10	815.2	−32.208	−19.941	3.8157	0.81042	2.1974	475.05
160	10	716.01	−9.9096	4.0567	3.9704	0.81369	2.6592	379.65
170	10	577.01	17.335	34.665	4.1557	0.82712	3.4932	291.48
180	10	423.27	46.729	70.354	4.3597	0.81609	3.3117	250.86
190	10	329.85	68.239	98.556	4.5124	0.78026	2.3941	250.55
200	10	277.64	83.705	119.72	4.6211	0.75064	1.8945	259
210	10	243.97	96.206	137.19	4.7064	0.72977	1.6246	268.66
220	10	219.88	107.09	152.56	4.7779	0.71517	1.4618	278.16
230	10	201.46	116.97	166.61	4.8404	0.70479	1.3546	287.21
240	10	186.72	126.21	179.76	4.8964	0.6973	1.2795	295.78
250	10	174.54	134.97	192.27	4.9474	0.69183	1.2244	303.91
260	10	164.23	143.41	204.29	4.9946	0.68785	1.1826	311.63
270	10	155.35	151.58	215.95	5.0386	0.68498	1.1499	319.01
280	10	147.59	159.56	227.32	5.0799	0.68298	1.124	326.06
290	10	140.71	167.38	238.45	5.119	0.68167	1.1031	332.84
300	10	134.57	175.08	249.39	5.1561	0.68095	1.086	339.35
400	10	95.666	248.94	353.47	5.456	0.69226	1.0168	394.54
500	10	75.32	322.06	454.83	5.6822	0.71874	1.0154	438.67
600	10	62.443	396.95	557.09	5.8686	0.74774	1.0312	476.82
700	10	53.456	474.1	661.17	6.029	0.77428	1.0504	511.24
800	10	46.788	553.42	767.15	6.1704	0.79695	1.0687	543.06
900	10	41.629	634.62	874.83	6.2973	0.81586	1.0847	572.92
1000	10	37.511	717.42	984	6.4123	0.83158	1.0984	601.18
50.0 MPa								
59.881	50	1334.9	−190.65	−153.2	2.1372	1.1052	1.5979	1253.3
60	50	1334.5	−190.47	−153.01	2.1404	1.1051	1.5988	1252.8
70	50	1297.7	−175.38	−136.85	2.3894	1.0837	1.6192	1200.6
80	50	1261.5	−160.38	−120.75	2.6044	1.0454	1.5995	1143.8
90	50	1225.8	−145.65	−104.86	2.7916	1.0065	1.578	1089.3
100	50	1190.4	−131.17	−89.167	2.9569	0.97167	1.5618	1037.3
110	50	1155.1	−116.9	−73.609	3.1052	0.94123	1.5505	987.55
120	50	1119.7	−102.8	−58.145	3.2398	0.91455	1.5429	939.95
130	50	1084.3	−88.857	−42.744	3.3631	0.89098	1.5377	894.5
140	50	1048.7	−75.064	−27.387	3.4769	0.87001	1.5339	851.29
150	50	1013.1	−61.419	−12.064	3.5826	0.85127	1.5308	810.46
160	50	977.39	−47.927	3.2301	3.6813	0.8345	1.5279	772.17
170	50	941.79	−34.598	18.493	3.7738	0.81947	1.5245	736.59
180	50	906.41	−21.446	33.717	3.8608	0.806	1.5201	703.87
190	50	871.4	−8.4887	48.89	3.9429	0.79393	1.5142	674.12

(continued)

Table 12.6 (continued)

Temperature (K)	Pressure (MPa)	Density (kg/m³)	Internal energy (kJ/kg)	Enthalpy (kJ/kg)	Entropy (kJ/kg K)	C_v (kJ/kg K)	C_p (kJ/kg K)	Sound speed (m/s)
200	50	836.95	4.2543	63.995	4.0204	0.78312	1.5063	647.41
210	50	803.28	16.763	79.008	4.0936	0.77344	1.4959	623.74
220	50	770.58	29.018	93.904	4.1629	0.76479	1.483	603.05
230	50	739.04	41.003	108.66	4.2285	0.75708	1.4675	585.22
240	50	708.83	52.707	123.25	4.2906	0.75022	1.4496	570.09
250	50	680.07	64.122	137.64	4.3494	0.74414	1.4298	557.42
260	50	652.83	75.249	151.84	4.405	0.73881	1.4087	546.97
270	50	627.16	86.092	165.82	4.4578	0.73416	1.3869	538.49
280	50	603.05	96.663	179.57	4.5078	0.73016	1.3648	531.73
290	50	580.46	106.97	193.11	4.5553	0.72676	1.343	526.44
300	50	559.33	117.04	206.44	4.6005	0.72392	1.3219	522.42
400	50	409.96	208.29	330.26	4.9577	0.71852	1.1746	520.51
500	50	326.44	291.27	444.44	5.2128	0.73575	1.1191	543.03
600	50	273.18	372.36	555.39	5.4151	0.75939	1.1038	569.97
700	50	235.91	453.75	665.7	5.5851	0.78259	1.1039	597.32
800	50	208.16	536.14	776.34	5.7329	0.80307	1.1095	624.1
900	50	186.57	619.66	887.65	5.864	0.82047	1.1168	650.04
1000	50	169.24	704.27	999.71	5.982	0.83513	1.1243	675.1

Reference

1. R. Schmidt, W. Wagner, A new form of the equation of state for pure substances and its applications to oxygen. Fluid Phase Equilib. **19**, 175 (1985)

Chapter 13
Methane

13.1 Formulation for Methane

The formulation for methane was published in 1991 by Setzmann and Wagner [1]. Because methane is the most abundant component in natural gas, the formulation is more often applied in mixture models with nitrogen, carbon dioxide, ethane, propane, and the butanes than as a pure fluid, and high accuracy is required, for example, in custody transfer in pipe lines. The equation of state is now more than 25 years old, and was one of the first to use the new Gaussian-bell shaped terms described in Chap. 2. There are only a few new high-accuracy measurements available for methane, but the fitting technology has changed significantly over 25 years, and the equation for the fluid properties is slated to be refit over the next two years with modern technology. The new equation will not change the calculated properties significantly due to the low uncertainties already in the equation, but will vastly improve the extrapolation properties and the shape of isotherms within the two-phase region. The behavior of the current equation in these regions has been a source of problems for mixture models. Because of the low uncertainties in the current equation in the single-phase region, the calculated properties given in this work will continue to be nearly the same as those from any new equation to be published in the future. The uncertainty statement in REFPROP is as follows.

The uncertainties in density are 0.03% for pressures below 12 MPa and temperatures below 350 K and up to 0.07% for pressures less than 50 MPa. For the speed of sound, the uncertainty ranges from 0.03% (in the vapor phase) to 0.3% depending on temperature and pressure. Heat capacities may be generally calculated within an uncertainty of 1% (Tables 13.1, 13.2, 13.3 and 13.4).

J.W. Leachman et al., *Thermodynamic Properties of Cryogenic Fluids*,
International Cryogenics Monograph Series, DOI 10.1007/978-3-319-57835-4_13

Table 13.1 Fixed point properties and correlation limits for methane

	Temperature (K)	Pressure (kPa)	Density (mol/dm^3)
Critical point	190.564	4599.2	10.139
Triple point	90.6941	11.696	28.142
Upper limit	625	1000,000	40.072
Lower limit	90.6941		

Table 13.2 Fluid constants and normal boiling point for methane

Molar mass	16.0428 g/mol
Ideal-gas constant	8.31451 J/(mol K)
Pitzer's acentric factor	0.01142
Normal boiling point	111.667 K

Table 13.3 Fundamental equation of state coefficients and exponents for methane

i	N_i	t_i	d_i	p_i	φ_i	β_i	γ_i	D_i
1	0.0436790102800	−0.5	1					
2	0.6709236199000	0.5	1					
3	−1.7655778590000	1	1					
4	0.8582330241000	0.5	2					
5	−1.2065130520000	1	2					
6	0.5120467220000	1.5	2					
7	−0.0004000010791	4.5	2					
8	−0.0124784242300	0	3					
9	0.0310026970100	1	4					
10	0.0017547485220	3	4					
11	−0.0000031719216050	1	8					
12	−0.0000022403468400	3	9					
13	0.0000002947056156	3	10					
14	0.1830487909000	0	1	1				
15	0.1511883679000	1	1	1				
16	−0.4289363877000	2	1	1				
17	0.0689400244600	0	2	1				
18	−0.0140831399600	0	4	1				
19	−0.0306305483000	2	5	1				
20	−0.0296990670800	2	6	1				
21	−0.0193204083100	5	1	2				
22	−0.1105739959000	5	2	2				
23	0.0995254899500	5	3	2				
24	0.0085484378250	2	4	2				
25	−0.0615055566200	4	4	2				
26	−0.0429179242300	12	3	3				
27	−0.0181320729000	8	5	3				

(continued)

Table 13.3 (continued)

i	N_i	t_i	d_i	p_i	φ_i	β_i	γ_i	D_i
28	0.0344590476000	10	5	3				
29	−0.0023859194500	10	8	3				
30	−0.0115909493900	10	2	4				
31	0.0664169360200	14	3	4				
32	−0.0237154959000	12	4	4				
33	−0.0396162490500	18	4	4				
34	−0.0138729204400	22	4	4				
35	0.0338948959900	18	5	4				
36	−0.0029273787530	14	6	4				
37	0.0000932479994600	2	2		20	200	1.07	1
38	−6.2871715180000	0	0		40	250	1.11	1
39	12.7106946700000	1	0		40	250	1.11	1
40	−6.4239534660000	2	0		40	250	1.11	1

Table 13.4 Ancillary function coefficients and exponents for methane

	α^0	
i	a_i	k_i
0	4.0016	
1	−2.9705496668	
2	2.8907453831	
3	0.008449	648.0
4	4.6942	1957.0
5	3.4865	3895.0
6	1.6572	5705.0
7	1.4115	15,080.0

$n = 7$ (there are no polynomial terms in the first summation)

	Vapor pressure		Saturated liquid density		Saturated vapor density	
i	N_i	k_i	N_i	k_i	N_i	k_i
1	−6.036219	1	1.990639	0.354	−1.88028	0.354
2	1.409353	1.5	−0.78756	0.5	−2.85265	2.5/3
3	−0.4945199	2	0.036977	2.5	−3.00065	4.5/3
4	−1.443048	4.5			−5.25117	7.5/3
5					−13.1919	12.5/3
6					−37.554	23.5/3

The functional form of the ancillary equations for methane are different than those of the other fluids, see Chap. 2 for the correct form

13.2 Property Plots for Methane

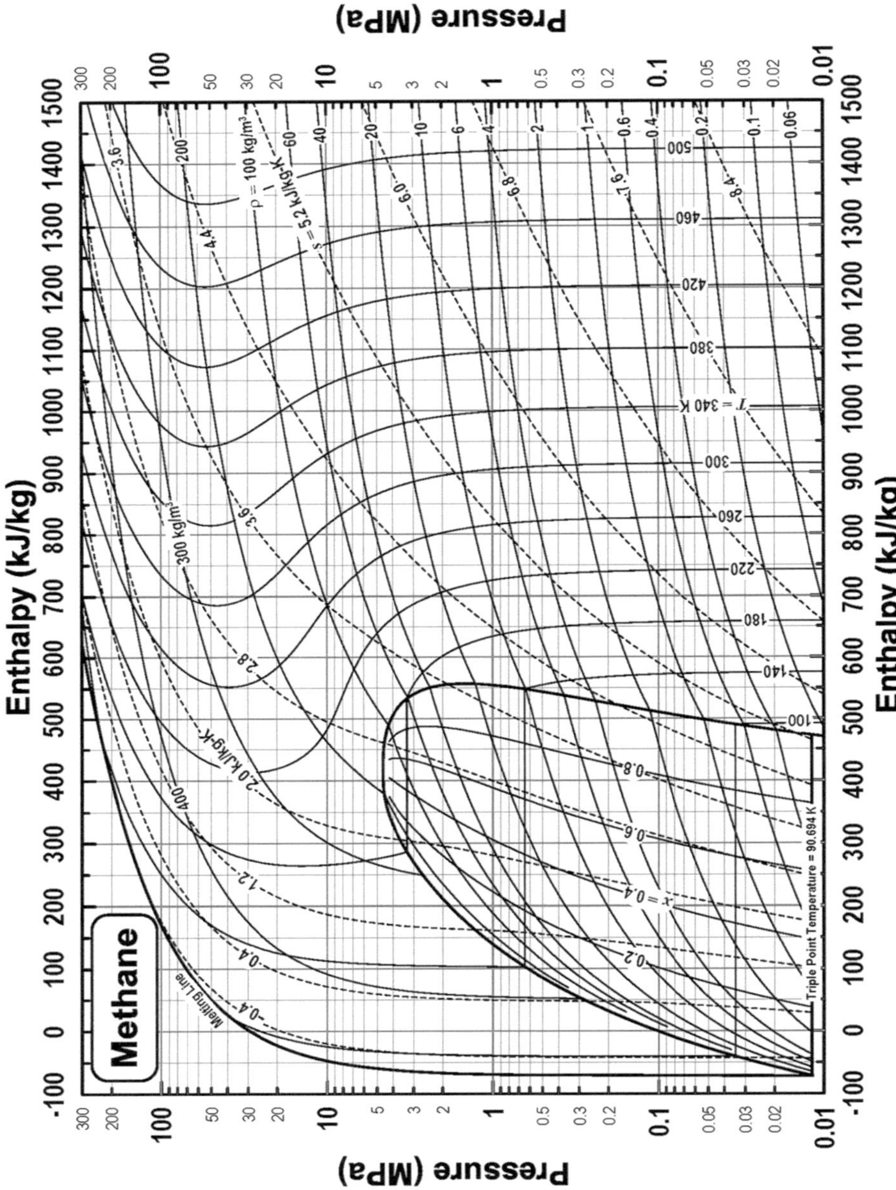
Methane
Enthalpy (kJ/kg)
Pressure (MPa)
Melting Line
Triple Point Temperature = 90.694 K
T = 340 K
ρ = 100 kg/m³
s = 5.2 kJ/kg-K
300 kg/m³
2.0 kJ/kg-K
x = 0.4

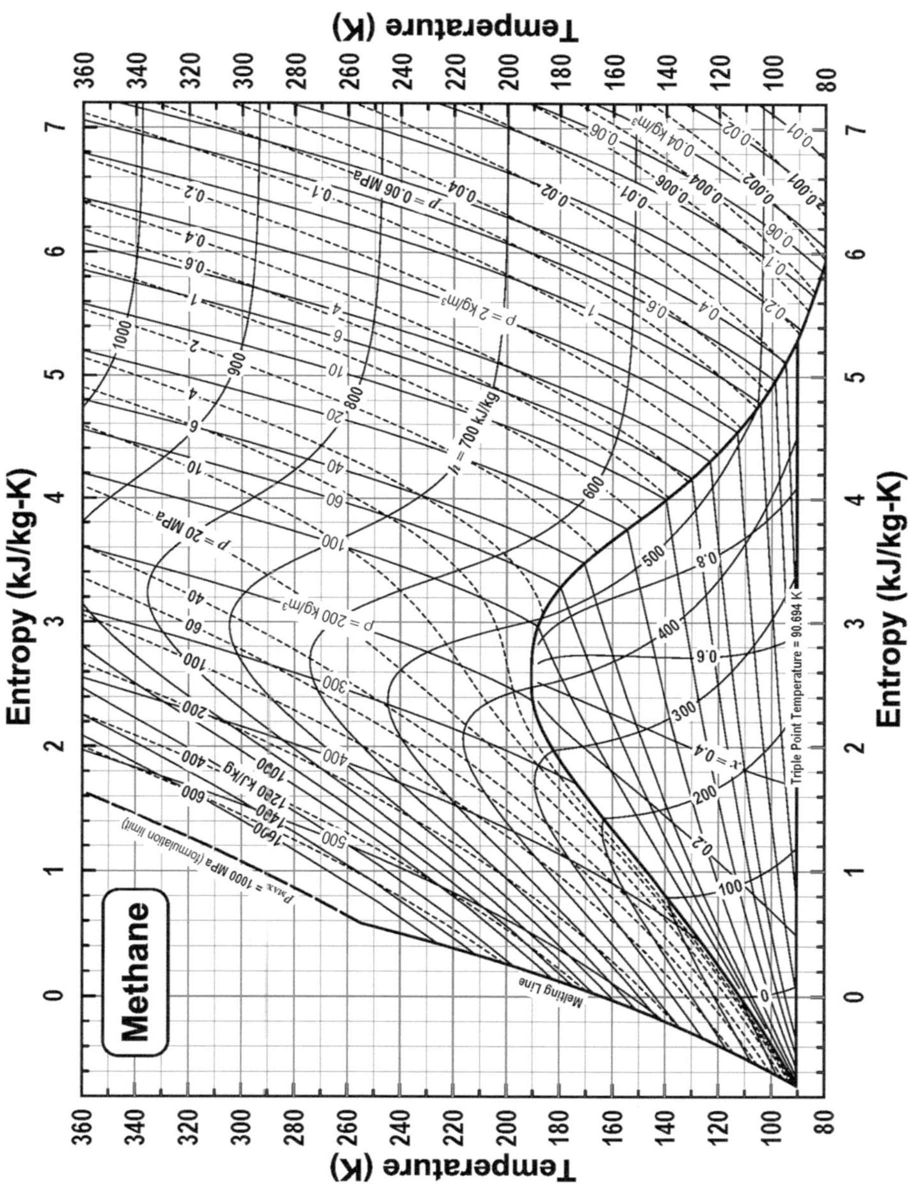
Methane
Entropy (kJ/kg-K)
Temperature (K)
Melting Line
Triple Point Temperature = 90.694 K
h = 700 kJ/kg
p = 20 MPa
p = 0.06 MPa
ρ = 200 kg/m³
ρ = 2 kg/m³
0.04 kg/m³
x = 0.4
1200 kJ/kg
p_{MAX} = 1000 MPa (formulation limit)

13.3 Thermodynamic Data Tables for Methane

See Tables 13.5 and 13.6.

Table 13.5 Thermodynamic properties of saturated methane

Temperature (K)	Pressure (MPa)	Liquid density (kg/m^3)	Vapor density (kg/m^3)	Liquid Enthalpy (kJ/kg)	Vapor enthalpy (kJ/kg)	Liquid entropy (kJ/kg K)	Vapor entropy (kJ/kg K)	Liquid C_v (kJ/kg K)	Vapor C_v (kJ/kg K)	Liquid sound speed (m/s)	Vapor sound speed (m/s)
90.694	0.011696	451.48	0.25074	−71.82	472.44	−0.70994	5.2911	2.1677	1.5735	1538.6	249.13
94	0.017613	447.05	0.3653	−60.657	478.86	−0.58919	5.1504	2.1471	1.5783	1508.4	253.2
98	0.027778	441.63	0.55488	−47.086	486.48	−0.44804	4.9965	2.1243	1.585	1471	257.87
102	0.04216	436.12	0.81329	−33.429	493.9	−0.31179	4.8581	2.1032	1.5926	1432.9	262.24
106	0.061868	430.5	1.1556	−19.675	501.09	−0.17997	4.7329	2.0832	1.6012	1394.1	266.29
110	0.08813	424.78	1.5982	−5.813	508.02	−0.05217	4.6191	2.0642	1.6108	1354.7	270.01
111.67	0.10132	422.36	1.8164	0	510.83	0	4.5746	2.0564	1.6151	1338.1	271.46
114	0.12228	418.93	2.1586	8.1709	514.67	0.071976	4.5149	2.0458	1.6214	1314.7	273.39
118	0.16574	412.95	2.8557	22.29	520.99	0.1928	4.419	2.0281	1.6329	1274	276.4
122	0.22002	406.82	3.7099	36.56	526.95	0.31062	4.3302	2.0112	1.6454	1232.7	279.03
126	0.28667	400.52	4.7434	51	532.52	0.42575	4.2474	1.995	1.6591	1190.8	281.28
130	0.36732	394.04	5.9804	65.629	537.67	0.53846	4.1695	1.9795	1.6739	1148.1	283.13
134	0.46363	387.35	7.4483	80.472	542.33	0.64904	4.0958	1.965	1.6901	1104.6	284.57
138	0.5773	380.42	9.1779	95.558	546.48	0.7578	4.0254	1.9515	1.7077	1060.2	285.58
142	0.71006	373.24	11.205	110.92	550.06	0.86501	3.9575	1.9391	1.7272	1014.9	286.16
146	0.86368	365.74	13.571	126.6	553	0.97101	3.8915	1.9281	1.7487	968.46	286.3
150	1.04	357.9	16.328	142.64	555.23	1.0761	3.8267	1.9186	1.7727	920.85	285.97
154	1.2408	349.64	19.537	159.12	556.66	1.1808	3.7622	1.9109	1.7997	871.87	285.16

(continued)

Table 13.5 (continued)

Temperature (K)	Pressure (MPa)	Liquid density (kg/m^3)	Vapor density (kg/m^3)	Liquid Enthalpy (kJ/kg)	Vapor enthalpy (kJ/kg)	Liquid entropy (kJ/kg K)	Vapor entropy (kJ/kg K)	Liquid C_v (kJ/kg K)	Vapor C_v (kJ/kg K)	Liquid sound speed (m/s)	Vapor sound speed (m/s)
158	1.468	340.9	23.279	176.09	557.19	1.2854	3.6974	1.9055	1.8303	821.35	283.86
162	1.7235	331.57	27.66	193.67	556.66	1.3905	3.6311	1.9028	1.8655	769.03	282.03
166	2.0096	321.5	32.822	211.99	554.87	1.4968	3.5624	1.9037	1.9066	714.59	279.65
170	2.3283	310.5	38.974	231.24	551.54	1.6054	3.4895	1.9095	1.9556	657.52	276.66
174	2.6822	298.25	46.434	251.72	546.23	1.7177	3.4103	1.9225	2.0157	597.05	273.02
178	3.074	284.21	55.74	273.93	538.23	1.8362	3.3211	1.9473	2.0925	531.94	268.6
182	3.5071	267.33	67.94	298.8	526.13	1.9657	3.2147	1.9944	2.1977	459.94	263.17
186	3.986	244.93	85.704	328.75	506.47	2.1182	3.0737	2.0978	2.362	375.88	255.97
190.56	4.5992	162.66	162.66	415.59	415.59	2.5624	2.5624				

Table 13.6 Thermodynamic properties of methane

Temperature (K)	Pressure (MPa)	Density (kg/m^3)	Internal energy (kJ/kg)	Enthalpy (kJ/kg)	Entropy (kJ/kg K)	C_v (kJ/kg K)	C_p (kJ/kg K)	Sound speed (m/s)
0.1 MPa								
90.717	0.1	451.5	−71.822	−71.601	−0.70968	2.1678	3.3674	1539
100	0.1	438.94	−40.394	−40.166	−0.3798	2.1138	3.4079	1452.6
110	0.1	424.79	−6.031	−5.7956	−0.05226	2.0642	3.4691	1354.8
111.51	0.1	422.59	−0.79398	−0.55734	−0.00497	2.0572	3.4799	1339.7
111.51	0.1	1.7946	454.84	510.56	4.5787	1.6147	2.2165	271.33
120	0.1	1.655	468.76	529.19	4.7397	1.5937	2.1741	282.84
130	0.1	1.5182	484.9	550.77	4.9125	1.581	2.1459	295.56
140	0.1	1.4034	500.88	572.14	5.0709	1.5739	2.1287	307.59
150	0.1	1.3055	516.77	593.37	5.2174	1.5699	2.1176	319.04
160	0.1	1.2208	532.59	614.51	5.3538	1.5679	2.1105	330.01
170	0.1	1.1467	548.38	635.59	5.4816	1.5675	2.1061	340.55
180	0.1	1.0813	564.15	656.63	5.6019	1.5684	2.104	350.71
190	0.1	1.023	579.93	677.67	5.7156	1.5707	2.104	360.51
200	0.1	0.97088	595.72	698.72	5.8236	1.5745	2.1058	369.98
210	0.1	0.92384	611.55	719.8	5.9264	1.5799	2.1096	379.14
220	0.1	0.88121	627.44	740.92	6.0247	1.5869	2.1154	388.01
230	0.1	0.84238	643.4	762.11	6.1189	1.5959	2.1232	396.6
240	0.1	0.80686	659.46	783.39	6.2095	1.6067	2.1331	404.91
250	0.1	0.77425	675.62	804.78	6.2968	1.6195	2.1451	412.96
260	0.1	0.74418	691.93	826.3	6.3812	1.6343	2.1592	420.77
270	0.1	0.71638	708.38	847.97	6.463	1.651	2.1754	428.34
280	0.1	0.6906	725.01	869.82	6.5424	1.6698	2.1936	435.68
290	0.1	0.66662	741.84	891.85	6.6197	1.6904	2.2138	442.81
300	0.1	0.64425	758.88	914.1	6.6951	1.7128	2.2358	449.74
400	0.1	0.4826	944.4	1151.6	7.3763	2.0134	2.534	510.56
500	0.1	0.38591	1164	1423.1	7.9807	2.3809	2.9005	561.86
600	0.1	0.32153	1420.9	1731.9	8.5426	2.7538	3.273	608.04
0.2 MPa								
90.742	0.2	451.53	−71.795	−71.352	−0.70938	2.1679	3.3669	1539.4
100	0.2	439.01	−40.465	−40.01	−0.38052	2.114	3.4071	1453.3
110	0.2	424.88	−6.1194	−5.6486	−0.05307	2.0645	3.468	1355.8
120.62	0.2	408.95	31.137	31.626	0.27035	2.0169	3.5551	1247
120.62	0.2	3.3966	466.06	524.94	4.3601	1.641	2.2996	278.17
130	0.2	3.1112	481.86	546.14	4.5294	1.6095	2.2304	291.19
140	0.2	2.8611	498.32	568.23	4.6931	1.593	2.1897	304.04
150	0.2	2.6516	514.56	589.98	4.8432	1.5834	2.1639	316.11
160	0.2	2.4727	530.65	611.53	4.9823	1.5778	2.1468	327.55
170	0.2	2.3176	546.64	632.94	5.1121	1.575	2.1355	338.47
180	0.2	2.1818	562.59	654.26	5.2339	1.5743	2.1283	348.93
190	0.2	2.0615	578.5	675.52	5.3489	1.5754	2.1244	358.98
200	0.2	1.9543	594.41	696.75	5.4578	1.5784	2.1233	368.67
210	0.2	1.8579	610.35	717.99	5.5614	1.5831	2.1248	378.01

(continued)

Table 13.6 (continued)

Temperature (K)	Pressure (MPa)	Density (kg/m^3)	Internal energy (kJ/kg)	Enthalpy (kJ/kg)	Entropy (kJ/kg K)	C_v (kJ/kg K)	C_p (kJ/kg K)	Sound speed (m/s)
220	0.2	1.7709	626.32	739.26	5.6604	1.5897	2.1286	387.02
230	0.2	1.6918	642.36	760.57	5.7551	1.5982	2.1349	395.74
240	0.2	1.6196	658.48	781.96	5.8461	1.6087	2.1435	404.16
250	0.2	1.5534	674.7	803.45	5.9338	1.6212	2.1544	412.31
260	0.2	1.4925	691.06	825.06	6.0186	1.6358	2.1676	420.2
270	0.2	1.4363	707.56	846.81	6.1007	1.6524	2.1829	427.85
280	0.2	1.3842	724.24	868.72	6.1804	1.671	2.2005	435.26
290	0.2	1.3358	741.1	890.82	6.2579	1.6915	2.22	442.44
300	0.2	1.2907	758.18	913.13	6.3336	1.7138	2.2415	449.43
400	0.2	0.96564	943.92	1151	7.0159	2.0138	2.5368	510.55
500	0.2	0.77183	1163.6	1422.8	7.6207	2.3811	2.9022	561.98
600	0.2	0.64295	1420.6	1731.6	8.1829	2.754	3.2741	608.22
0.5 MPa								
90.819	0.5	451.63	−71.714	−70.607	−0.70849	2.1681	3.3655	1540.8
100	0.5	439.24	−40.679	−39.541	−0.38266	2.1148	3.4048	1455.7
110	0.5	425.15	−6.3834	−5.2074	−0.05547	2.0653	3.4647	1358.5
120	0.5	410.25	28.599	29.818	0.24924	2.0205	3.5443	1256.9
130	0.5	394.22	64.511	65.78	0.53703	1.9799	3.6546	1149.9
135.35	0.5	385.04	84.241	85.54	0.68597	1.9603	3.7335	1089.7
135.35	0.5	8.0016	481.31	543.79	4.0716	1.6959	2.5127	284.96
140	0.5	7.6355	489.79	555.28	4.1551	1.6645	2.4334	292.28
150	0.5	6.9757	507.39	579.07	4.3192	1.6302	2.3348	306.66
160	0.5	6.4398	524.46	602.1	4.4679	1.6114	2.2756	319.78
170	0.5	5.9914	541.19	624.65	4.6046	1.6	2.2364	331.98
180	0.5	5.6083	557.71	646.87	4.7316	1.5935	2.2097	343.45
190	0.5	5.2758	574.1	668.87	4.8506	1.5906	2.1916	354.32
200	0.5	4.9837	590.39	690.72	4.9627	1.5906	2.1798	364.67
210	0.5	4.7245	606.65	712.48	5.0688	1.5932	2.173	374.56
220	0.5	4.4925	622.9	734.19	5.1698	1.5981	2.1704	384.05
230	0.5	4.2835	639.17	755.9	5.2663	1.6054	2.1714	393.16
240	0.5	4.094	655.5	777.63	5.3588	1.6149	2.1757	401.93
250	0.5	3.9212	671.91	799.42	5.4478	1.6266	2.1831	410.38
260	0.5	3.763	688.43	821.3	5.5336	1.6405	2.1932	418.52
270	0.5	3.6175	705.08	843.3	5.6166	1.6565	2.2061	426.39
280	0.5	3.4832	721.89	865.43	5.6971	1.6747	2.2214	434
290	0.5	3.3588	738.87	887.73	5.7754	1.6948	2.2391	441.37
300	0.5	3.2432	756.05	910.22	5.8516	1.7168	2.259	448.5
400	0.5	2.4174	942.51	1149.3	6.5375	2.015	2.5451	510.54
500	0.5	1.9297	1162.6	1421.7	7.1437	2.3817	2.907	562.35
600	0.5	1.6066	1419.8	1731	7.7066	2.7544	3.2772	608.78
1.0 MPa								
90.947	1	451.79	−71.579	−69.366	−0.70701	2.1686	3.3632	1543
100	1	439.62	−41.033	−38.758	−0.38621	2.1162	3.4011	1459.6

(continued)

Table 13.6 (continued)

Temperature (K)	Pressure (MPa)	Density (kg/m^3)	Internal energy (kJ/kg)	Enthalpy (kJ/kg)	Entropy (kJ/kg K)	C_v (kJ/kg K)	C_p (kJ/kg K)	Sound speed (m/s)
110	1	425.61	−6.8197	−4.4701	−0.05946	2.0667	3.4593	1363.2
120	1	410.8	28.055	30.489	0.24468	2.0219	3.5363	1262.4
130	1	394.92	63.82	66.352	0.53168	1.9813	3.6424	1156.6
140	1	377.51	100.85	103.5	0.80692	1.946	3.7984	1043.7
149.14	1	359.62	136.38	139.16	1.0536	1.9205	4.0227	931.21
149.14	1	15.698	491.11	554.81	3.8406	1.7673	2.8765	286.08
150	1	15.536	492.91	557.27	3.8571	1.7556	2.8402	287.76
160	1	13.965	512.62	584.23	4.0311	1.6824	2.5878	305.18
170	1	12.769	531.1	609.42	4.1838	1.6499	2.4604	320.2
180	1	11.807	548.9	633.59	4.322	1.6307	2.3803	333.74
190	1	11.007	566.26	657.11	4.4492	1.6191	2.3264	346.2
200	1	10.326	583.33	680.18	4.5675	1.6131	2.2895	357.81
210	1	9.7351	600.22	702.94	4.6786	1.6113	2.2643	368.73
220	1	9.2166	616.99	725.49	4.7835	1.6131	2.2477	379.06
230	1	8.7563	633.71	747.91	4.8832	1.6179	2.2379	388.88
240	1	8.344	650.42	770.27	4.9783	1.6255	2.2336	398.24
250	1	7.9721	667.16	792.6	5.0695	1.6358	2.234	407.2
260	1	7.6344	683.97	814.96	5.1572	1.6485	2.2385	415.79
270	1	7.326	700.88	837.38	5.2418	1.6635	2.2465	424.04
280	1	7.0432	717.92	859.9	5.3237	1.6809	2.2578	431.98
290	1	6.7826	735.11	882.55	5.4032	1.7003	2.272	439.64
300	1	6.5415	752.48	905.35	5.4805	1.7217	2.2889	447.04
400	1	4.8457	940.15	1146.5	6.1724	2.0171	2.5592	510.57
500	1	3.8595	1160.8	1419.9	6.781	2.3828	2.9151	562.99
600	1	3.2104	1418.4	1729.9	7.3451	2.755	3.2824	609.73
2.0 MPa								
91.201	2	452.1	−71.308	−66.884	−0.70405	2.1694	3.3586	1547.5
100	2	440.37	−41.731	−37.189	−0.39325	2.1188	3.3937	1467.4
110	2	426.5	−7.6782	−2.9889	−0.06733	2.0695	3.4489	1372.3
120	2	411.88	26.988	31.844	0.23571	2.0248	3.5211	1273.3
130	2	396.27	62.472	67.519	0.52121	1.9841	3.6193	1169.7
140	2	379.27	99.098	104.37	0.79426	1.9484	3.7606	1060.1
150	2	360.21	137.44	142.99	1.0606	1.9197	3.9813	941.45
160	2	337.78	178.63	184.55	1.3287	1.9028	4.3754	807.65
165.87	2	321.84	205.18	211.39	1.4934	1.9036	4.8055	716.36
165.87	2	32.644	493.68	554.95	3.5646	1.9052	3.9235	279.73
170	2	30.533	504.6	570.1	3.6549	1.8197	3.4665	290.52
180	2	26.904	527.55	601.88	3.8367	1.7313	2.9678	311.34
190	2	24.354	548.15	630.27	3.9902	1.6906	2.7325	328.4
200	2	22.392	567.52	656.84	4.1265	1.6669	2.5923	343.3
210	2	20.805	586.15	682.28	4.2506	1.6532	2.5009	356.7
220	2	19.48	604.29	706.96	4.3655	1.6465	2.439	368.98
230	2	18.348	622.12	731.12	4.4729	1.6452	2.3965	380.37

(continued)

Table 13.6 (continued)

Temperature (K)	Pressure (MPa)	Density (kg/m³)	Internal energy (kJ/kg)	Enthalpy (kJ/kg)	Entropy (kJ/kg K)	C_v (kJ/kg K)	C_p (kJ/kg K)	Sound speed (m/s)
240	2	17.364	639.75	754.93	4.5742	1.6483	2.3678	391.03
250	2	16.497	657.27	778.51	4.6705	1.655	2.3493	401.07
260	2	15.724	674.75	801.95	4.7624	1.665	2.3388	410.57
270	2	15.03	692.25	825.31	4.8506	1.6779	2.3348	419.61
280	2	14.402	709.8	848.66	4.9355	1.6935	2.3363	428.22
290	2	13.83	727.44	872.05	5.0176	1.7115	2.3423	436.47
300	2	13.307	745.22	895.52	5.0971	1.7317	2.3522	444.38
400	2	9.7326	935.42	1140.9	5.8015	2.0212	2.5875	510.82
500	2	7.7187	1157.4	1416.5	6.4149	2.385	2.9312	564.39
600	2	6.4093	1415.6	1727.7	6.9813	2.7562	3.2927	611.69
5.0 MPa								
91.963	5	453.03	−70.482	−59.445	−0.69522	2.172	3.3454	1560.7
100	5	442.55	−43.752	−32.454	−0.41385	2.1266	3.3733	1490
110	5	429.09	−10.149	1.5037	−0.09024	2.0777	3.4202	1398.6
120	5	415	23.945	35.993	0.20982	2.0332	3.4802	1304.3
130	5	400.1	58.672	71.169	0.49134	1.9925	3.5587	1206.8
140	5	384.13	94.248	107.26	0.75878	1.9561	3.6663	1105.3
150	5	366.57	131.02	144.65	1.0167	1.9251	3.8215	998.65
160	5	347.04	169.56	183.97	1.2704	1.9015	4.0616	884.35
170	5	323.91	211.02	226.46	1.5278	1.8898	4.4792	758.12
180	5	293.96	258.14	275.15	1.8059	1.9037	5.4115	609.53
190	5	240.69	324.26	345.04	2.1826	2.0517	10.712	393.44
200	5	87.764	487.43	544.4	3.2099	1.9965	7.2726	291.29
210	5	69.687	527.13	598.88	3.4762	1.8404	4.4121	319.48
220	5	60.479	555.72	638.39	3.6602	1.7789	3.6049	340.2
230	5	54.337	580.28	672.29	3.8109	1.7457	3.2141	357.45
240	5	49.769	602.74	703.2	3.9425	1.7277	2.9851	372.5
250	5	46.159	623.94	732.26	4.0612	1.7196	2.8379	385.97
260	5	43.193	644.36	760.12	4.1704	1.7187	2.7385	398.24
270	5	40.689	664.25	787.14	4.2724	1.7234	2.6699	409.56
280	5	38.532	683.82	813.58	4.3686	1.7326	2.6226	420.08
290	5	36.644	703.19	839.64	4.46	1.7455	2.5906	429.95
300	5	34.972	722.46	865.43	4.5475	1.7616	2.5703	439.25
400	5	24.598	921.21	1124.5	5.2919	2.033	2.6743	513.11
500	5	19.273	1147	1406.5	5.9199	2.3912	2.9788	569.49
600	5	15.927	1407.5	1721.5	6.4932	2.7599	3.3229	618.13
10.0 MPa								
93.222	10	454.54	−69.075	−47.075	−0.68061	2.1761	3.3251	1582.3
100	10	446.02	−46.899	−24.478	−0.44663	2.1389	3.3437	1525.7
110	10	433.13	−13.956	9.1319	−0.12632	2.0907	3.3797	1439.8
120	10	419.79	19.323	43.144	0.1696	2.0467	3.4243	1351.9
130	10	405.87	53.015	77.654	0.44579	2.0061	3.4799	1262.3
140	10	391.22	87.234	112.79	0.70618	1.9693	3.5516	1171

(continued)

Table 13.6 (continued)

Temperature (K)	Pressure (MPa)	Density (kg/m^3)	Internal energy (kJ/kg)	Enthalpy (kJ/kg)	Entropy (kJ/kg K)	C_v (kJ/kg K)	C_p (kJ/kg K)	Sound speed (m/s)
150	10	375.63	122.14	148.76	0.9543	1.9366	3.6467	1077.6
160	10	358.81	157.97	185.84	1.1936	1.9089	3.7765	981.8
170	10	340.33	195.08	224.47	1.4277	1.8872	3.959	882.98
180	10	319.56	234.01	265.3	1.661	1.873	4.2262	780.76
190	10	295.43	275.62	309.47	1.8997	1.8688	4.6379	675.09
200	10	266.19	321.34	358.9	2.1531	1.878	5.3036	567.92
210	10	229.53	373.1	416.67	2.4348	1.8981	6.2783	469.31
220	10	187.59	429.44	482.74	2.7421	1.9034	6.713	404.38
230	10	152.24	480.6	546.28	3.0246	1.8795	5.8612	382.48
240	10	128.4	521.74	599.62	3.2518	1.8467	4.8533	383.61
250	10	112.43	555.52	644.47	3.435	1.8199	4.1685	392.22
260	10	101.01	584.81	683.81	3.5893	1.8026	3.7301	402.84
270	10	92.361	611.29	719.56	3.7243	1.7941	3.4407	413.78
280	10	85.507	635.97	752.92	3.8456	1.793	3.2424	424.47
290	10	79.893	659.44	784.61	3.9568	1.7977	3.1025	434.73
300	10	75.175	682.08	815.1	4.0602	1.8074	3.0023	444.53
400	10	49.736	897.68	1098.7	4.8773	2.0509	2.8187	522.58
500	10	38.322	1130.2	1391.2	5.5288	2.4008	3.0548	580.99
600	10	31.474	1394.4	1712.2	6.1132	2.7658	3.3704	630.63
50.0 MPa								
102.89	50	465.34	−56.753	50.696	−0.57013	2.2052	3.2158	1737.1
110	50	458.14	−35.579	73.558	−0.35527	2.1748	3.2154	1693.5
120	50	448.03	−5.8904	105.71	−0.07551	2.1348	3.2149	1632.3
130	50	437.92	23.678	137.85	0.18178	2.0971	3.2138	1572.1
140	50	427.81	53.111	169.99	0.4199	2.0617	3.2125	1513.3
150	50	417.69	82.398	202.11	0.6415	2.0291	3.2115	1456.1
160	50	407.55	111.53	234.22	0.84875	1.9993	3.2111	1400.7
170	50	397.39	140.51	266.33	1.0434	1.9725	3.2116	1347.4
180	50	387.23	169.33	298.45	1.227	1.9488	3.213	1296.3
190	50	377.07	197.99	330.59	1.4008	1.9283	3.2153	1247.4
200	50	366.92	226.49	362.76	1.5658	1.911	3.2182	1201
210	50	356.8	254.82	394.96	1.7229	1.8969	3.2217	1157
220	50	346.74	282.99	427.19	1.8729	1.886	3.2253	1115.6
230	50	336.75	310.99	459.46	2.0163	1.8784	3.2288	1077
240	50	326.88	338.81	491.77	2.1538	1.8739	3.2319	1041
250	50	317.15	366.44	524.1	2.2858	1.8725	3.2342	1007.8
260	50	307.59	393.89	556.45	2.4127	1.8741	3.2356	977.4
270	50	298.23	421.15	588.81	2.5348	1.8786	3.236	949.68
280	50	289.11	448.22	621.17	2.6525	1.8858	3.2354	924.6
290	50	280.26	475.1	653.51	2.766	1.8956	3.2338	902.06
300	50	271.69	501.8	685.84	2.8756	1.9079	3.2314	881.94
400	50	203.43	762.26	1008.1	3.8026	2.1333	3.2276	779.36
500	50	161.43	1026.5	1336.2	4.5342	2.4583	3.3582	768.59

(continued)

Table 13.6 (continued)

Temperature (K)	Pressure (MPa)	Density (kg/m³)	Internal energy (kJ/kg)	Enthalpy (kJ/kg)	Entropy (kJ/kg K)	C_v (kJ/kg K)	C_p (kJ/kg K)	Sound speed (m/s)
600	50	134.47	1311.2	1683	5.1659	2.8062	3.5887	786.69
100.0 MPa								
114.13	100	476.63	−39.739	170.07	−0.44802	2.2377	3.1449	1897.8
120	100	471.72	−23.482	188.51	−0.29049	2.2163	3.1364	1870
130	100	463.48	4.0379	219.8	−0.04003	2.1811	3.1214	1823.5
140	100	455.37	31.329	250.93	0.19073	2.1477	3.1059	1778.4
150	100	447.37	58.385	281.91	0.40448	2.1163	3.0904	1734.7
160	100	439.48	85.201	312.74	0.60345	2.0872	3.0752	1692.7
170	100	431.7	111.78	343.42	0.78944	2.0608	3.0607	1652.3
180	100	424.04	138.13	373.96	0.964	2.037	3.0472	1613.5
190	100	416.48	164.26	404.37	1.1284	2.0162	3.0348	1576.5
200	100	409.03	190.18	434.66	1.2838	1.9983	3.0239	1541
210	100	401.69	215.9	464.85	1.4311	1.9834	3.0145	1507.1
220	100	394.47	241.45	494.96	1.5711	1.9715	3.0068	1474.8
230	100	387.36	266.83	524.99	1.7047	1.9628	3.0008	1444
240	100	380.37	292.08	554.98	1.8323	1.9571	2.9966	1414.6
250	100	373.51	317.2	584.93	1.9546	1.9544	2.9943	1386.7
260	100	366.78	342.22	614.87	2.072	1.9548	2.9938	1360.2
270	100	360.17	367.17	644.81	2.185	1.9579	2.9951	1335.1
280	100	353.7	392.05	674.78	2.294	1.9639	2.9982	1311.3
290	100	347.36	416.9	704.78	2.3992	1.9725	3.003	1288.8
300	100	341.16	441.73	734.84	2.5012	1.9836	3.0095	1267.5
400	100	286.86	693.18	1041.8	3.3832	2.1987	3.1493	1115.8
500	100	245.53	960.22	1367.5	4.1091	2.5128	3.3746	1044.8
600	100	214.28	1251.3	1718	4.7473	2.8501	3.638	1018.4

Reference

1. U. Setzmann, W. Wagner, A new equation of state and tables of thermodynamic properties for methane covering the range from the melting line to 625 K at Pressures up to 1000 MPa. J. Phys. Chem. Ref. Data **20**(6), 1061 (1991)

Chapter 14
Krypton

14.1 Formulation for Krypton

The formulation for krypton was published by Lemmon and Span [1]. The uncertainty statement provided in REFPROP is as follows.

The equation of state is valid from the triple point to 750 K with pressures to 200 MPa, although the uncertainties increase substantially above 100 MPa. The uncertainties in density are typically 0.2% below 100 MPa, increasing to 1% at pressures up to 200 MPa. The uncertainty in vapor pressure is 0.2% and the uncertainties in speed of sound are 0.01% in the vapor phase (including supercritical conditions) at low pressures, 1% below 20 MPa in the liquid phase, and 3% below 100 MPa at other state points. The limited amount of heat capacity data show that the uncertainty is 1% near the triple point, and uncertainties in heat capacities at other states are probably within 2%, at least at pressures up to 20 MPa (Tables 14.1, 14.2, 14.3 and 14.4).

J.W. Leachman et al., *Thermodynamic Properties of Cryogenic Fluids*,
International Cryogenics Monograph Series, DOI 10.1007/978-3-319-57835-4_14

Table 14.1 Fixed point properties and correlation limits for krypton

	Temperature (K)	Pressure (kPa)	Density (mol/dm^3)
Critical point	209.48	5525.0	10.85
Triple point	115.775	73.5	29.2
Upper limit	750	200,000	33.42
Lower limit	115.775		

Table 14.2 Fluid constants and normal boiling point for krypton

Molar mass	83.798 g/mol
Ideal-gas constant	8.314472 J/(mol K)
Pitzer's acentric factor	−0.00089
Normal boiling point	119.73 K

Table 14.3 Fundamental equation of state coefficients and exponents for krypton

i	N_i	t_i	d_i	p_i
1	0.83561	0.25	1	
2	−2.3725	1.125	1	
3	0.54567	1.5	1	
4	0.014361	1.375	2	
5	0.066502	0.25	3	
6	0.000193	0.875	7	
7	0.16818	0.625	2	1
8	−0.03313	1.75	5	1
9	−0.15008	3.625	1	2
10	−0.0229	3.625	4	2
11	−0.02145	14.5	3	3
12	0.00694	12	4	3

Table 14.4 Ancillary function coefficients and exponents for krypton

α^0	
i	a_i
0	2.5
1	−3.7506412806
2	3.7798018435
$m = 0$, $n = 0$	

	Vapor pressure		Saturated liquid density		Saturated vapor density	
i	N_i	k_i	N_i	k_i	N_i	k_i
1	−5.9697	1	20.593	0.62	−6.4163	0.525
2	1.2673	1.5	−65.49	0.84	8.9956	0.77
3	−0.95609	2.95	94.407	1.07	−10.216	1.04
4	−35.63	9.3	−69.678	1.34	−13.477	3.2
5	56.884	10.4	22.81	1.6	−211.52	8.3
6					213.75	9

14.2 Property Plots for Krypton

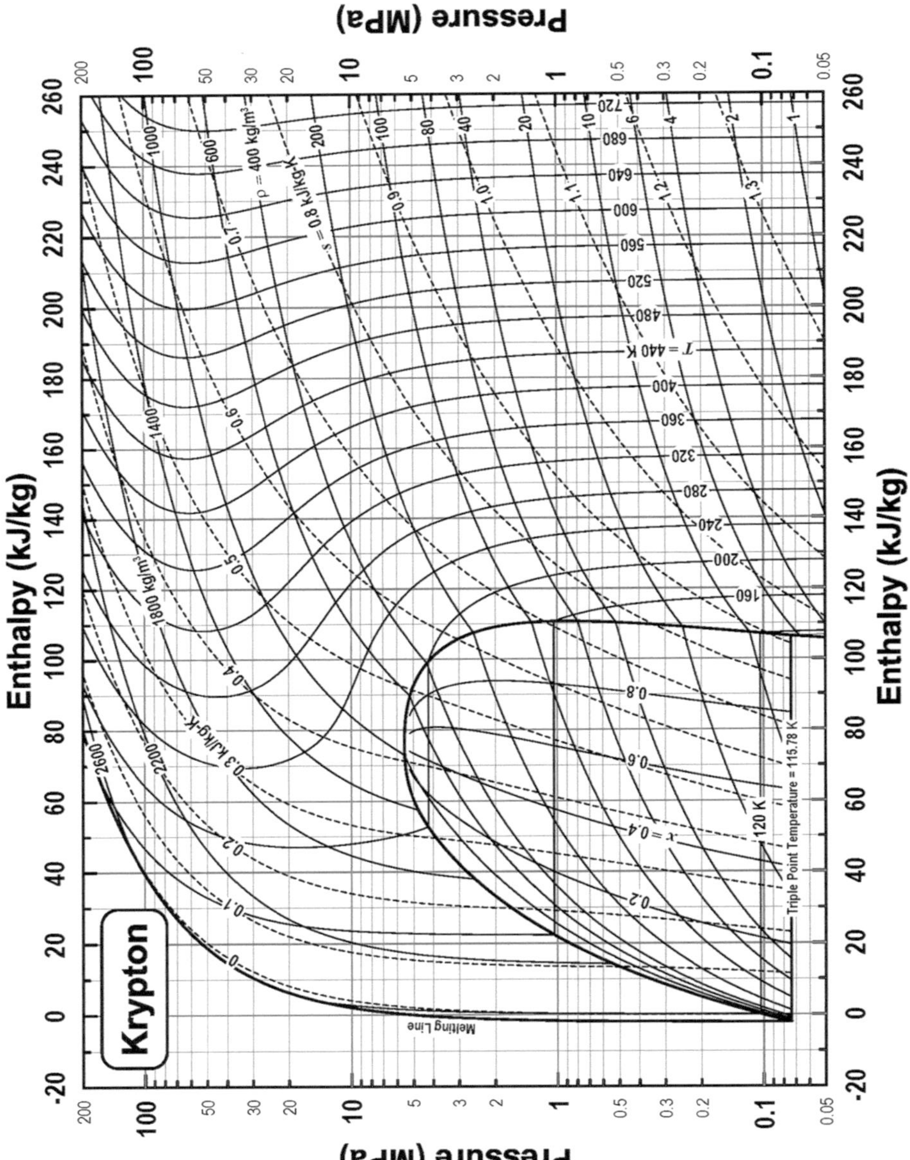
Krypton
Enthalpy (kJ/kg)
Pressure (MPa)
Melting Line
Triple Point Temperature = 115.78 K

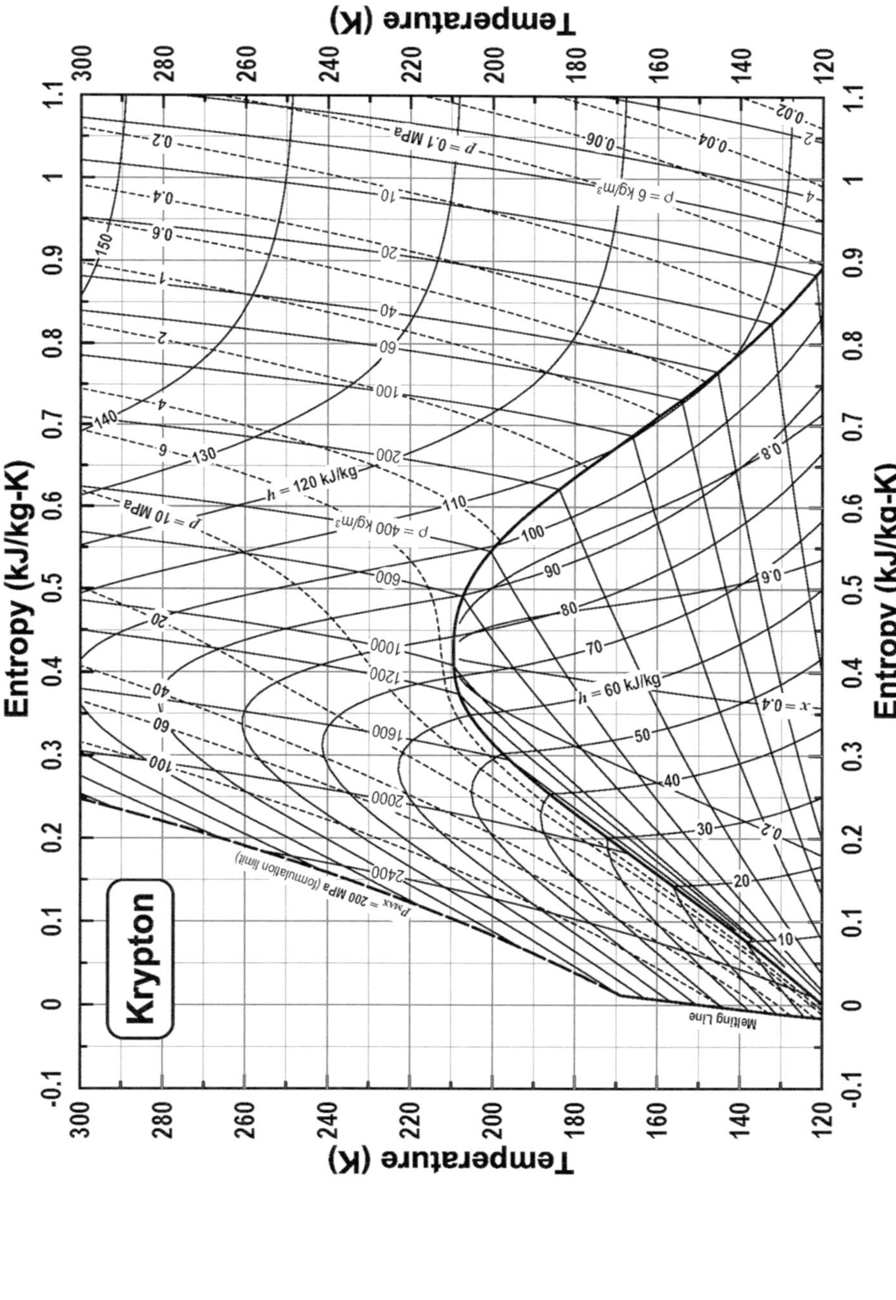
Krypton
Entropy (kJ/kg-K)
Temperature (K)
Melting Line
p_{MAX} = 200 MPa (formulation limit)
h = 120 kJ/kg
h = 60 kJ/kg
p = 10 MPa
p = 0.1 MPa
ρ = 400 kg/m³
ρ = 6 kg/m³
x = 0.4

14.3 Thermodynamic Data Tables for Krypton

See Tables 14.5 and 14.6.

Table 14.5 Thermodynamic properties of saturated krypton

Temperature (K)	Pressure (MPa)	Liquid density (kg/m^3)	Vapor density (kg/m^3)	Liquid enthalpy (kJ/kg)	Vapor enthalpy (kJ/kg)	Liquid entropy (kJ/kg K)	Vapor entropy (kJ/kg K)	Liquid C_v (kJ/kg K)	Vapor C_v (kJ/kg K)	Liquid sound speed (m/s)	Vapor sound speed (m/s)
115.78	0.073534	2446.6	6.5738	−2.0601	106.34	−0.0174	0.91888	0.25329	0.15836	698.95	136.13
119.73	0.10133	2416.6	8.8182	0	107.05	0	0.8941	0.24953	0.1604	682.54	137.89
124	0.13966	2383.7	11.837	2.2343	107.77	0.018204	0.86931	0.24572	0.16286	664.72	139.64
128	0.18487	2352.3	15.318	4.3477	108.39	0.034827	0.84764	0.24236	0.16542	647.85	141.15
132	0.24037	2320.2	19.514	6.4817	108.94	0.05106	0.82728	0.2392	0.1682	630.81	142.51
136	0.30754	2287.4	24.517	8.6399	109.43	0.06695	0.80808	0.23621	0.17122	613.55	143.74
140	0.38778	2253.9	30.425	10.826	109.85	0.082536	0.78988	0.2334	0.17445	596.06	144.82
144	0.48254	2219.4	37.346	13.044	110.2	0.097859	0.77253	0.23075	0.17789	578.31	145.77
148	0.59329	2184	45.398	15.299	110.46	0.11296	0.75593	0.22827	0.18154	560.25	146.57
152	0.72152	2147.5	54.718	17.595	110.63	0.12787	0.73995	0.22596	0.18539	541.83	147.23
156	0.86874	2109.8	65.459	19.938	110.71	0.14264	0.72448	0.22382	0.18945	523.02	147.75
160	1.0365	2070.7	77.8	22.335	110.67	0.1573	0.70941	0.22189	0.19374	503.74	148.13
164	1.2263	2030	91.957	24.793	110.52	0.1719	0.69464	0.22017	0.19827	483.91	148.36
168	1.4398	1987.5	108.19	27.322	110.24	0.1865	0.68006	0.21871	0.20308	463.45	148.44
172	1.6786	1942.8	126.81	29.933	109.81	0.20114	0.66554	0.21756	0.20824	442.24	148.37
176	1.9444	1895.7	148.25	32.639	109.21	0.21589	0.65097	0.21678	0.2138	420.13	148.14
180	2.239	1845.6	173.04	35.458	108.42	0.23085	0.63617	0.2165	0.21989	396.95	147.75
184	2.5642	1791.8	201.91	38.416	107.39	0.24612	0.62095	0.21688	0.22666	372.44	147.18
188	2.9221	1733.3	235.93	41.547	106.07	0.26186	0.60506	0.21818	0.23435	346.3	146.43
192	3.3149	1668.6	276.7	44.902	104.39	0.2783	0.58812	0.22083	0.24332	318.08	145.46
196	3.7452	1595.1	326.87	48.568	102.21	0.29583	0.56954	0.22557	0.25419	287.14	144.22
200	4.2163	1507.9	391.41	52.706	99.309	0.3152	0.54821	0.23392	0.26809	252.47	142.62
204	4.7323	1395.1	482.09	57.712	95.127	0.33821	0.52161	0.2496	0.2875	212.21	140.39
208	5.3	1202.5	648.81	65.386	87.424	0.37332	0.47927	0.28839	0.32056	161.23	136.48
209.48	5.5255	909.21	909.21	76.278	76.278	0.42445	0.42445				

Table 14.6 Thermodynamic properties of krypton

Temperature (K)	Pressure (MPa)	Density (kg/m³)	Internal energy (kJ/kg)	Enthalpy (kJ/kg)	Entropy (kJ/kg K)	C_v (kJ/kg K)	C_p (kJ/kg K)	Sound speed (m/s)
0.1 MPa								
115.78	0.1	2446.7	−2.0925	−2.0516	−0.01742	0.25329	0.51709	699.04
119.57	0.1	2417.9	−0.12917	−0.08781	−0.00073	0.24968	0.51969	683.24
119.57	0.1	8.7124	95.547	107.02	0.89511	0.16031	0.2746	137.82
130	0.1	7.9507	97.262	109.84	0.91769	0.15602	0.26589	144.52
140	0.1	7.3446	98.857	112.47	0.9372	0.1538	0.26112	150.5
150	0.1	6.8291	100.42	115.07	0.95511	0.15241	0.25802	156.16
160	0.1	6.3841	101.97	117.64	0.97169	0.15148	0.25588	161.56
170	0.1	5.9954	103.51	120.19	0.98715	0.15084	0.25435	166.76
180	0.1	5.6526	105.03	122.72	1.0017	0.15039	0.25321	171.77
190	0.1	5.3476	106.55	125.25	1.0153	0.15005	0.25235	176.62
200	0.1	5.0745	108.07	127.77	1.0282	0.14981	0.25169	181.32
210	0.1	4.8284	109.58	130.29	1.0405	0.14962	0.25116	185.89
220	0.1	4.6054	111.08	132.8	1.0522	0.14948	0.25074	190.35
230	0.1	4.4022	112.59	135.3	1.0633	0.14937	0.25041	194.69
240	0.1	4.2165	114.09	137.8	1.074	0.14928	0.25013	198.94
250	0.1	4.0459	115.59	140.3	1.0842	0.14921	0.2499	203.09
260	0.1	3.8887	117.09	142.8	1.094	0.14916	0.2497	207.16
270	0.1	3.7434	118.58	145.3	1.1034	0.14911	0.24954	211.14
280	0.1	3.6086	120.08	147.79	1.1125	0.14907	0.2494	215.05
290	0.1	3.4832	121.58	150.29	1.1212	0.14904	0.24928	218.88
300	0.1	3.3663	123.07	152.78	1.1297	0.14902	0.24917	222.65
400	0.1	2.5213	138	177.66	1.2013	0.14891	0.2486	257.24
500	0.1	2.0161	152.91	202.51	1.2567	0.14887	0.24838	287.64
600	0.1	1.6797	167.81	227.34	1.302	0.14886	0.24827	315.11
700	0.1	1.4396	182.7	252.16	1.3402	0.14885	0.2482	340.36
0.2 MPa								
115.81	0.2	2445.9	−2.0915	−2.0098	−0.01741	0.2533	0.51696	699.29
120	0.2	2415	0.080252	0.16307	0.00102	0.24931	0.51986	681.83
129.17	0.2	2343	4.8852	4.9706	0.039621	0.24142	0.52918	642.88
129.17	0.2	16.469	96.412	108.56	0.84154	0.16621	0.29056	141.56
130	0.2	16.345	96.56	108.8	0.8434	0.16547	0.28905	142.14
140	0.2	15.006	98.288	111.62	0.8643	0.15947	0.27632	148.72
150	0.2	13.893	99.945	114.34	0.8831	0.15626	0.26905	154.76
160	0.2	12.947	101.56	117.01	0.9003	0.15426	0.26433	160.43
170	0.2	12.13	103.14	119.63	0.91622	0.15292	0.26103	165.82
180	0.2	11.416	104.71	122.23	0.93107	0.15198	0.25863	170.99
190	0.2	10.784	106.26	124.81	0.945	0.1513	0.25683	175.97
200	0.2	10.221	107.8	127.37	0.95814	0.1508	0.25545	180.77
210	0.2	9.7162	109.33	129.92	0.97058	0.15042	0.25437	185.43
220	0.2	9.2599	110.86	132.46	0.98239	0.15013	0.25351	189.96
230	0.2	8.8456	112.38	134.99	0.99364	0.14991	0.25282	194.36
240	0.2	8.4676	113.89	137.51	1.0044	0.14973	0.25225	198.66
250	0.2	8.1211	115.41	140.03	1.0147	0.14959	0.25178	202.86

(continued)

Table 14.6 (continued)

Temperature (K)	Pressure (MPa)	Density (kg/m^3)	Internal energy (kJ/kg)	Enthalpy (kJ/kg)	Entropy (kJ/kg K)	C_v (kJ/kg K)	C_p (kJ/kg K)	Sound speed (m/s)
260	0.2	7.8023	116.91	142.55	1.0245	0.14948	0.25138	206.96
270	0.2	7.508	118.42	145.06	1.034	0.14939	0.25105	210.98
280	0.2	7.2353	119.93	147.57	1.0431	0.14932	0.25076	214.91
290	0.2	6.9821	121.43	150.08	1.0519	0.14926	0.25051	218.77
300	0.2	6.7461	122.93	152.58	1.0604	0.14921	0.2503	222.56
400	0.2	5.046	137.91	177.54	1.1323	0.14898	0.24915	257.28
500	0.2	4.0328	152.84	202.43	1.1878	0.14892	0.2487	287.74
600	0.2	3.3592	167.75	227.29	1.2331	0.14889	0.24848	315.23
700	0.2	2.8787	182.65	252.13	1.2714	0.14888	0.24835	340.49
0.5 MPa								
115.9	0.5	2447.5	−2.0885	−1.8842	−0.01739	0.25331	0.51657	700.07
120	0.5	2416.4	0.033028	0.23995	0.000625	0.24941	0.51934	683.04
130	0.5	2337.9	5.2661	5.4799	0.042562	0.24085	0.5295	640.7
140	0.5	2254.6	10.627	10.849	0.082345	0.23343	0.54546	596.68
144.67	0.5	2213.6	13.195	13.42	0.10041	0.23032	0.55552	575.3
144.67	0.5	38.616	97.3	110.25	0.7697	0.17849	0.3297	145.91
150	0.5	36.783	98.364	111.96	0.78131	0.17145	0.31343	149.96
160	0.5	33.877	100.23	114.99	0.80089	0.16393	0.29505	156.74
170	0.5	31.473	102	117.88	0.81842	0.15973	0.28411	162.87
180	0.5	29.433	103.7	120.69	0.83444	0.15704	0.27679	168.57
190	0.5	27.669	105.35	123.43	0.84926	0.1552	0.27155	173.96
200	0.5	26.124	106.98	126.12	0.86309	0.15387	0.26764	179.1
210	0.5	24.755	108.58	128.78	0.87607	0.15288	0.26464	184.03
220	0.5	23.533	110.17	131.42	0.88832	0.15214	0.26229	188.78
230	0.5	22.432	111.74	134.03	0.89994	0.15156	0.26041	193.37
240	0.5	21.435	113.3	136.62	0.91099	0.15112	0.25888	197.83
250	0.5	20.527	114.85	139.21	0.92153	0.15076	0.25763	202.16
260	0.5	19.696	116.39	141.78	0.93161	0.15048	0.25658	206.38
270	0.5	18.932	117.93	144.34	0.94128	0.15025	0.25571	210.49
280	0.5	18.227	119.46	146.89	0.95057	0.15007	0.25496	214.52
290	0.5	17.574	120.99	149.44	0.9595	0.14991	0.25432	218.46
300	0.5	16.968	122.51	151.98	0.96811	0.14979	0.25377	222.31
400	0.5	12.64	137.62	177.18	1.0406	0.14921	0.2508	257.43
500	0.5	10.087	152.63	202.2	1.0965	0.14905	0.24967	288.03
600	0.5	8.3967	167.59	227.13	1.1419	0.14899	0.24912	315.59
700	0.5	7.1935	182.52	252.03	1.1803	0.14895	0.24881	340.87
1.0 MPa								
116.05	1	2448.4	−2.0834	−1.675	−0.01734	0.25333	0.51592	701.36
120	1	2418.6	−0.04509	0.36837	−2.8E−05	0.24958	0.5185	685.05
130	1	2340.5	5.171	5.5982	0.041828	0.24101	0.52829	643
140	1	2257.9	10.51	10.953	0.081502	0.23358	0.54367	599.38
150	1	2168.7	16.036	16.497	0.11975	0.22717	0.56688	553.34
159.17	1	2078.9	21.353	21.834	0.15427	0.22227	0.59908	507.77
159.17	1	75.104	97.374	110.69	0.7125	0.19283	0.39	148.06

(continued)

Table 14.6 (continued)

Temperature (K)	Pressure (MPa)	Density (kg/m³)	Internal energy (kJ/kg)	Enthalpy (kJ/kg)	Entropy (kJ/kg K)	C_v (kJ/kg K)	C_p (kJ/kg K)	Sound speed (m/s)
160	1	74.449	97.577	111.01	0.71451	0.19079	0.38439	148.84
170	1	67.682	99.827	114.6	0.7363	0.17493	0.33971	157.12
180	1	62.387	101.84	117.87	0.755	0.16712	0.3163	164.12
190	1	58.042	103.73	120.96	0.77168	0.16249	0.3017	170.39
200	1	54.372	105.53	123.92	0.78689	0.15942	0.29165	176.18
210	1	51.209	107.27	126.8	0.80093	0.15725	0.28432	181.62
220	1	48.443	108.97	129.61	0.81402	0.15565	0.27875	186.78
230	1	45.995	110.64	132.38	0.82631	0.15444	0.27442	191.71
240	1	43.807	112.28	135.1	0.83792	0.1535	0.27097	196.44
250	1	41.836	113.9	137.8	0.84892	0.15277	0.26817	201.01
260	1	40.05	115.5	140.47	0.85939	0.15218	0.26587	205.43
270	1	38.421	117.09	143.12	0.86939	0.15171	0.26395	209.72
280	1	36.928	118.67	145.75	0.87896	0.15133	0.26234	213.89
290	1	35.553	120.24	148.37	0.88814	0.15102	0.26096	217.96
300	1	34.283	121.8	150.97	0.89697	0.15076	0.25979	221.94
400	1	25.364	137.15	176.58	0.97068	0.1496	0.25359	257.69
500	1	20.19	152.28	201.81	1.027	0.14928	0.2513	288.54
600	1	16.789	167.31	226.88	1.0727	0.14914	0.25019	316.19
700	1	14.376	182.3	251.86	1.1112	0.14907	0.24956	341.52
2.0 MPa								
116.35	2	2450.3	−2.0727	−1.2565	−0.01725	0.25337	0.51465	703.91
120	2	2423	−0.19919	0.62624	−0.00132	0.24991	0.51686	689.02
130	2	2345.8	4.9839	5.8365	0.040378	0.24133	0.52595	647.55
140	2	2264.3	10.279	11.163	0.079842	0.23388	0.54025	604.69
150	2	2176.8	15.746	16.665	0.1178	0.22742	0.56162	559.71
160	2	2080.8	21.47	22.431	0.155	0.22199	0.59398	511.42
170	2	1971.8	27.596	28.61	0.19244	0.21798	0.64655	457.64
176.79	2	1885.1	32.123	33.183	0.21881	0.21668	0.70603	415.67
176.79	2	152.84	95.986	109.07	0.64808	0.21495	0.53558	148.08
180	2	146.34	97.043	110.71	0.65727	0.20338	0.48683	151.97
190	2	130.88	99.825	115.11	0.68105	0.18319	0.40379	161.76
200	2	119.58	102.19	118.92	0.70062	0.17329	0.36269	169.62
210	2	110.67	104.34	122.41	0.71767	0.16741	0.33783	176.45
220	2	103.34	106.35	125.7	0.73297	0.16349	0.32104	182.63
230	2	97.135	108.26	128.85	0.74696	0.16069	0.30893	188.34
240	2	91.781	110.1	131.89	0.75991	0.15861	0.2998	193.71
250	2	87.091	111.89	134.85	0.772	0.15701	0.29269	198.79
260	2	82.931	113.63	137.75	0.78336	0.15576	0.28703	203.64
270	2	79.207	115.34	140.6	0.79411	0.15476	0.28243	208.29
280	2	75.845	117.03	143.4	0.80431	0.15396	0.27864	212.77
290	2	72.79	118.69	146.17	0.81403	0.15331	0.27547	217.1
300	2	69.998	120.34	148.91	0.82332	0.15277	0.27279	221.3
400	2	51.052	136.2	175.38	0.89955	0.15037	0.2593	258.3
500	2	40.438	151.58	201.04	0.95683	0.14972	0.25456	289.59

(continued)

Table 14.6 (continued)

Temperature (K)	Pressure (MPa)	Density (kg/m^3)	Internal energy (kJ/kg)	Enthalpy (kJ/kg)	Entropy (kJ/kg K)	C_v (kJ/kg K)	C_p (kJ/kg K)	Sound speed (m/s)
600	2	33.556	166.77	226.37	1.003	0.14945	0.25231	317.43
700	2	28.705	181.86	251.53	1.0418	0.14931	0.25104	342.83
5.0 MPa								
117.25	5	2455.8	−2.0366	−0.00065	−0.01697	0.25351	0.511	711.42
120	5	2435.7	−0.64523	1.4075	−0.0051	0.25091	0.51232	700.55
130	5	2360.9	4.4459	6.5637	0.036166	0.24231	0.51957	660.66
140	5	2282.6	9.6226	11.813	0.075063	0.23481	0.53108	619.82
150	5	2199.3	14.93	17.203	0.11224	0.22823	0.54795	577.58
160	5	2109.6	20.427	22.797	0.14834	0.22254	0.57236	533.23
170	5	2010.6	26.203	28.69	0.18405	0.21783	0.60871	485.71
180	5	1897.6	32.408	35.043	0.22035	0.21453	0.66715	433.05
190	5	1760.5	39.359	42.199	0.25901	0.21406	0.77788	371.02
200	5	1566.6	48.041	51.233	0.30528	0.22308	1.1077	286.26
205.94	5	1321	57.001	60.787	0.35225	0.26296	3.2288	189.55
205.94	5	545.11	83.027	92.2	0.50478	0.30067	4.1473	138.88
210	5	434.83	89.043	100.54	0.54497	0.24289	1.3048	152.38
220	5	347.1	95.142	109.55	0.58697	0.20186	0.68573	168
230	5	303.02	99.005	115.51	0.61348	0.18611	0.52905	178.1
240	5	273.39	102.1	120.39	0.63426	0.17742	0.45429	186.18
250	5	251.2	104.79	124.69	0.65183	0.17179	0.40981	193.17
260	5	233.55	107.22	128.63	0.66729	0.16778	0.38012	199.46
270	5	218.99	109.49	132.32	0.68122	0.16479	0.35884	205.25
280	5	206.64	111.63	135.82	0.69397	0.16246	0.34286	210.65
290	5	195.96	113.67	139.19	0.70577	0.16062	0.33043	215.75
300	5	186.6	115.65	142.44	0.7168	0.15913	0.32051	220.6
400	5	129.9	133.32	171.81	0.80158	0.15275	0.2773	260.8
500	5	101.42	149.5	198.8	0.86185	0.15108	0.26438	293.14
600	5	83.67	165.15	224.91	0.90948	0.15039	0.25856	321.36
700	5	71.388	180.56	250.59	0.94907	0.15003	0.2554	346.89
10.0 MPa								
118.74	10	2464.7	−1.9643	2.093	−0.01645	0.25375	0.50541	723.45
120	10	2455.8	−1.3398	2.7321	−0.0111	0.25257	0.50579	718.71
130	10	2384.5	3.6178	7.8116	0.029557	0.24395	0.51062	681.03
140	10	2310.5	8.627	12.955	0.067671	0.23641	0.51865	642.94
150	10	2233	13.719	18.197	0.10383	0.22974	0.53032	604.18
160	10	2151.1	18.928	23.576	0.13854	0.22384	0.54647	564.46
170	10	2063.4	24.3	29.146	0.1723	0.21867	0.56859	523.4
180	10	1967.9	29.895	34.977	0.20562	0.21426	0.59935	480.49
190	10	1861.8	35.807	41.178	0.23914	0.21078	0.64376	435.04
200	10	1740	42.184	47.931	0.27376	0.20865	0.71222	386.03
210	10	1593.3	49.304	55.58	0.31106	0.20882	0.82925	332.15
220	10	1402.2	57.748	64.879	0.35428	0.21319	1.0577	272.84
230	10	1135.5	68.491	77.297	0.40942	0.2211	1.4212	218.22
240	10	861.1	79.696	91.309	0.46908	0.21404	1.2595	197.84

(continued)

Table 14.6 (continued)

Temperature (K)	Pressure (MPa)	Density (kg/m^3)	Internal energy (kJ/kg)	Enthalpy (kJ/kg)	Entropy (kJ/kg K)	C_v (kJ/kg K)	C_p (kJ/kg K)	Sound speed (m/s)
250	10	694.64	87.554	101.95	0.51257	0.19941	0.89668	198.89
260	10	596.43	93.039	109.81	0.5434	0.18903	0.6954	203.83
270	10	530.79	97.306	116.15	0.56734	0.18193	0.58234	209.35
280	10	482.8	100.88	121.59	0.58716	0.17676	0.51174	214.83
290	10	445.56	104.01	126.46	0.60423	0.17282	0.46387	220.13
300	10	415.47	106.84	130.91	0.61934	0.1697	0.42941	225.22
400	10	265.96	128.45	166.05	0.72121	0.15671	0.30908	267.23
500	10	203.2	146.08	195.29	0.78657	0.15332	0.28051	300.18
600	10	166.27	162.53	222.67	0.83653	0.15195	0.26858	328.56
700	10	141.37	178.45	249.19	0.87741	0.15122	0.26228	354.04
50.0 MPa								
130.25	50	2524.4	−1.0056	18.801	−0.011	0.25588	0.47458	805.32
140	50	2470.5	3.1773	23.416	0.023172	0.24853	0.4727	778.61
150	50	2415.1	7.4341	28.137	0.055747	0.24183	0.47172	751.91
160	50	2359.4	11.661	32.853	0.08618	0.23584	0.47148	725.92
170	50	2303.5	15.863	37.569	0.11477	0.23043	0.47178	700.64
180	50	2247.3	20.041	42.29	0.14175	0.22552	0.47246	676.11
190	50	2190.8	24.196	47.019	0.16732	0.22103	0.47337	652.37
200	50	2134.1	28.328	51.757	0.19163	0.21691	0.47436	629.47
210	50	2077.1	32.433	56.506	0.21479	0.21312	0.4753	607.47
220	50	2020	36.51	61.263	0.23692	0.20963	0.47605	586.43
230	50	1962.9	40.553	66.026	0.2581	0.20639	0.47647	566.43
240	50	1905.9	44.557	70.791	0.27838	0.2034	0.47645	547.53
250	50	1849.3	48.515	75.553	0.29782	0.20062	0.47587	529.8
260	50	1793.2	52.422	80.306	0.31646	0.19804	0.47464	513.28
270	50	1737.8	56.271	85.043	0.33434	0.19564	0.4727	498.02
280	50	1683.4	60.055	89.757	0.35148	0.19339	0.47003	484.03
290	50	1630.1	63.768	94.441	0.36792	0.1913	0.46663	471.31
300	50	1578.2	67.405	99.088	0.38367	0.18933	0.46253	459.84
400	50	1156.6	99.221	142.45	0.50885	0.17508	0.40186	400.68
500	50	899.05	124.23	179.85	0.59251	0.16709	0.34982	395.62
600	50	738.63	145.41	213.1	0.65323	0.16248	0.31808	407
700	50	630.27	164.54	243.87	0.7007	0.15963	0.29882	423.14
100.0 MPa								
143.82	100	2582.7	0.72838	39.447	−0.00317	0.25822	0.45319	887.4
150	100	2555.1	3.1012	42.239	0.015836	0.25414	0.45077	874.34
160	100	2510.8	6.9006	46.729	0.044815	0.24809	0.44733	853.83
170	100	2467	10.651	51.187	0.071843	0.24265	0.44433	834.07
180	100	2423.6	14.355	55.617	0.097163	0.23771	0.44165	815.06
190	100	2380.7	18.016	60.021	0.12097	0.23321	0.43918	796.8
200	100	2338.2	21.633	64.401	0.14344	0.22908	0.43685	779.27
210	100	2296.2	25.208	68.758	0.1647	0.22528	0.43458	762.47
220	100	2254.7	28.74	73.092	0.18487	0.22176	0.43233	746.41
230	100	2213.7	32.231	77.404	0.20403	0.2185	0.43004	731.07

(continued)

Table 14.6 (continued)

Temperature (K)	Pressure (MPa)	Density (kg/m^3)	Internal energy (kJ/kg)	Enthalpy (kJ/kg)	Entropy (kJ/kg K)	C_v (kJ/kg K)	C_p (kJ/kg K)	Sound speed (m/s)
240	100	2173.2	35.678	81.693	0.22229	0.21547	0.42769	716.47
250	100	2133.3	39.083	85.958	0.2397	0.21265	0.42525	702.58
260	100	2094	42.443	90.198	0.25633	0.21001	0.42271	689.4
270	100	2055.4	45.759	94.412	0.27223	0.20755	0.42007	676.92
280	100	2017.4	49.03	98.598	0.28746	0.20524	0.4173	665.14
290	100	1980.1	52.255	102.76	0.30205	0.20307	0.41443	654.04
300	100	1943.6	55.435	106.89	0.31605	0.20104	0.41146	643.59
400	100	1621.5	84.755	146.42	0.43002	0.18606	0.37894	570.49
500	100	1376.6	110.15	182.79	0.51128	0.1771	0.34941	537.49
600	100	1192.7	132.71	216.55	0.5729	0.17131	0.32695	525.89
700	100	1052.3	153.35	248.38	0.622	0.16732	0.31052	525.42

Reference

1. E.W. Lemmon, R. Span, Short fundamental equations of state for 20 industrial fluids. J. Chem. Eng. Data **51**, 785–850 (2006)

Index

A
Adiabatic bulk modulus, 19
Adiabatic compressibility, 19
Air
ancillary function coefficients and exponents, 119
fixed point properties and correlation limits, 118
fluid constants and normal boiling point, 118
fundamental equation of state coefficients and exponents, 118
pressure-enthalpy diagram, 119
saturation properties, 122
single-phase properties, 117
temperature-entropy diagram, 119
Ancillary functions, 18, 20, 26, 44, 58, 71, 76, 91, 105, 119, 133, 147, 159, 173, 187, 201
Argon
ancillary function coefficients and exponents, 159
fixed point properties and correlation limits, 158
fluid constants and normal boiling point, 118
pressure-enthalpy diagram, 159
saturation properties, 162
single-phase properties, 163
temperature-entropy diagram, 159

C
Carbon monoxide
ancillary function coefficients and exponents, 133
fixed point properties and correlation limits, 132
fluid constants and normal boiling point, 132
fundamental equation of state coefficients and exponents, 132
pressure-enthalpy diagram, 133
saturation properties, 136
single-phase properties, 137
temperature-entropy diagram, 133
Compressibility factor, 19
Criteria for equations of state, 4, 16
Critical region, 3, 6, 17, 24, 45, 57, 89, 103, 145, 157, 171

D
Data correlation, 2, 3, 5, 13
Deuterium
ancillary function coefficients and exponents, 76
fixed point properties and correlation limits, 75
fluid constants and normal boiling point, 75
pressure-enthalpy diagram, 76
saturation properties, 79
single-phase properties, 80
temperature-entropy diagram, 76

E
Enthalpy, 4, 9, 11, 19, 29, 43, 48, 49, 61, 62, 79, 80, 94, 108, 122, 136, 150, 190, 204
Entropy, 9, 11, 19, 23, 29, 43, 48, 49, 61, 79, 80, 94, 108, 118, 122, 136, 150, 162, 176, 190, 204
Estimated accuracies, 6, 14

F
Fit variables, 14
Fluorine
ancillary function coefficients and exponents, 147
fixed point properties and correlation limits, 146

J.W. Leachman et al., *Thermodynamic Properties of Cryogenic Fluids*,
International Cryogenics Monograph Series, DOI 10.1007/978-3-319-57835-4

Fluorine (*cont.*)
- fluid constants and normal boiling point, 146
- pressure-enthalpy diagram, 147
- saturation properties, 150
- single-phase properties, 151
- temperature-entropy diagram, 147

Fugacity coefficient, 19
Fundamental equation, 4, 11, 18, 24, 25, 43, 44, 57, 70, 75, 104, 118, 132, 146, 158, 172, 186, 200

G

Gibbs energy, 4, 19

H

Helium
- ancillary function coefficients and exponents, 26
- fixed point properties and correlation limits, 24
- fluid constants and normal boiling point, 25
- fundamental equation of state coefficients and exponents, 25
- pressure-enthalpy diagram, 26
- saturation properties, 23
- single-phase properties, 30
- temperature-entropy diagram, 29

Helmholtz energy, 4, 11–13, 16, 19, 74
Hydrogen
- ancillary function coefficients and exponents, 71
- fixed point properties and correlation limits, 43
- fluid constants and normal boiling point, 57
- fundamental equation of state coefficients and exponents, 57
- pressure-enthalpy diagram, 58
- saturation properties, 61
- single-phase properties, 62
- temperature-entropy diagram, 58

I

Internal energy, 4, 19, 49, 62, 80, 95, 109, 123, 137, 151, 163, 177, 192, 205
Isentropic expansion coefficient, 19
Isobaric heat capacity, 3, 6, 19
Isochoric heat capacity, 3, 6, 19
Isothermal bulk modulus, 19
Isothermal compressibility, 19
Isothermal expansion coefficien, 19

J

Joule-Thomson coefficient, 3, 19

K

Krypton
- ancillary function coefficients and exponents, 201
- fixed point properties and correlation limits, 200
- fluid constants and normal boiling point, 200
- fundamental equation of state coefficients and exponents, 200
- pressure-enthalpy diagram, 201
- saturation properties, 204
- single-phase properties, 205
- temperature-entropy diagram, 201

M

Methane
- ancillary function coefficients and exponents, 187
- fixed point properties and correlation limits, 186
- fluid constants and normal boiling point, 186
- fundamental equation of state coefficients and exponents, 186
- pressure-enthalpy diagram, 187
- saturation properties, 190
- single-phase properties, 191
- temperature-entropy diagram, 187

N

Neon
- ancillary function coefficients and exponents, 91
- fixed point properties and correlation limits, 90
- fluid constants and normal boiling point, 90
- fundamental equation of state coefficients and exponents, 108
- pressure-enthalpy diagram, 91
- saturation properties, 94
- single-phase properties, 95
- temperature-entropy diagram, 91

Nitrogen
- ancillary function coefficients and exponents, 105
- fixed point properties and correlation limits, 104
- fluid constants and normal boiling point, 104
- fundamental equation of state coefficients and exponents, 104
- pressure-enthalpy diagram, 105
- saturation properties, 108

single-phase properties, 109
temperature-entropy diagram, 105
Non-linear regression, 14, 15

O

Orthohydrogen
ancillary function coefficients and exponents, 71
fixed point properties and correlation limits, 70
fluid constants and normal boiling point, 70
fundamental equation of state coefficients and exponents, 43, 70
Oxygen
ancillary function coefficients and exponents, 173
fixed point properties and correlation limits, 172
fluid constants and normal boiling point, 172
fundamental equation of state coefficients and exponents, 172
pressure-enthalpy diagram, 173
saturation properties, 176
single-phase properties, 177
temperature-entropy diagram, 173

P

Parahydrogen
ancillary function coeficients and exponents, 44
fixed point properties and correlation limits, 43
fluid constants and normal boiling point, 44
fundamental equation of state coefficients and exponents, 43, 44
pressure-enthalpy diagram, 45
saturation properties, 48
single-phase properties, 49
temperature-entropy diagram, 45
Pressure, 2–4, 6–9, 11, 15–19, 24, 26, 29, 42–45, 48, 49, 57, 58, 61, 62, 70, 71, 74–76, 79, 80, 89–91, 94, 103–105, 108, 117–119, 122, 131–133, 136, 146, 147, 150, 157–159, 171–173, 176, 185–187, 190, 199–201, 204

Q

Quantum parameters, 7, 8

R

REFPROP, 8, 18, 24, 43, 57, 89, 131, 145, 171, 185, 199

S

Second virial coefficient, 19
Speed of sound, 3, 6, 16, 19, 24, 45, 57, 89, 103, 117, 131, 145, 171, 185, 199

T

Third virial coefficient, 3, 19

V

Volume expansivity, 19

MIX
Papier aus verantwortungsvollen Quellen
Paper from responsible sources
FSC® C105338

FSC
www.fsc.org

If you have any concerns about our products,
you can contact us on
ProductSafety@springernature.com

In case Publisher is established outside the EU,
the EU authorized representative is:
Springer Nature Customer Service Center GmbH
Europaplatz 3, 69115 Heidelberg, Germany

Printed by Libri Plureos GmbH
in Hamburg, Germany